8th International Conference on Fine Particle Magnetism

(ICFPM 2013)

Journal of Physics: Conference Series Volume 521

Perpignan, France
24-27 June 2013

Editor:

Hamid Kachkachi

ISBN: 978-1-63439-302-7
ISSN: 1742-6588

Printed from e-media with permission by:

Curran Associates, Inc.
57 Morehouse Lane
Red Hook, NY 12571

Some format issues inherent in the e-media version may also appear in this print version.

Printed by Curran Associates, Inc. (2014)

For permission requests, please contact the Institute of Physics at the address below.

Institute of Physics
Dirac House, Temple Back
Bristol BS1 6BE UK

Phone: 44 1 17 929 7481
Fax: 44 1 17 920 0979

techtracking@iop.org

Additional copies of this publication are available from:

Curran Associates, Inc.
57 Morehouse Lane
Red Hook, NY 12571 USA
Phone: 845-758-0400
Fax: 845-758-2634
Email: curran@proceedings.com
Web: www.proceedings.com

TABLE OF CONTENTS

STRUCTURAL, SURFACE AND FINITE-SIZE EFFECTS

ASSEMBLIES, COLLECTIVE EFFECTS

APPLICATIONS : DRUG DELIVERY, HYPETHERMIA, MAGNETO-RESISTANCE

Journal of Physics: Conference Series **521** (2014) 011001 doi:10.1088/1742-6596/521/1/011001

Preface

The 8th International Conference on Fine Particle Magnetism (ICFPM) was held in Perpignan from 24 to 27 June 2013, and was the continuation of the previous meetings held in Bangor (1996), Rome (1991), Barcelona (1999), Pittsburg (2002), London (2004), Rome (2007) and Uppsala (2010). The next meeting will be organized by Profs. Robert D. Shull, George Hadjipanayis and Cindi Dennis, in 2016 at NIST, Gaithersburg (USA).

ICFPM is a small-sized conference focused on the magnetism of nanoparticles. It provides an international forum for discussing the state-of-the-art understanding of physics of these systems, of their properties and the underlying phenomena, as approached from a variety of directions: theory and modelling, experiments on well characterized or model systems (both fabricated and synthetised), as well as experiments on technologically-relevant non-ideal systems.

This meeting brought together about 120 participants working on experimental, theoretical and applied topics of the multidisciplinary research areas covered by magnetic nanoparticles, with focused interest on either single-particle or collective phenomena. The technical program of the conference was based on keynote conferences, invited talks, oral contributions and poster sessions, covering the following aspects:

- Fabrication, synthesis, characterization
- Single particle, surface and finite-size effects on magnetic properties
- Magnetization dynamics, micro-wave assisted switching, dynamical coupling
- Assemblies, collective effects, self-assembling and nanostructuring
- Applications : hyperthermia, drug delivery, magneto-caloric, magneto-resistance, magneto-plasmonics, magnetic particle imaging

This ICFPM edition was organized by the group Nanoscale Spin Systems of the laboratory PROMES of the CNRS (UPR8521), and Université de Perpignan Via Domitia. The meeting took place at the Congress Center of the city of Perpignan providing high-quality facilities for the technical program as well for the personal comfort of the participants. After a sight-seeing tour in the beautiful town of Collioure, on the Vermeille coast, the poster session took place in a XII century nice Château, followed by the conference gala dinner held at the same place.

We are extremely grateful to all colleagues who helped organize the conference and set up its scientific program, especially the Organizing Committee and the national and international Scientific Committees. We would also like to thank all of our colleagues who participated in the pre-selection of the contributions submitted to the conference and those who have acted as reviewers of the papers included in these proceedings.

Last but not least we would like to acknowledge a strong support from the director of our Lab, Prof. Gilles Flamant, and the vice-President of our university, Prof. Xavier Py, who both took part in the conference opening. We also thank la Région Languedoc-Roussillon for financial support.

Scientific context

Magnetic nanoparticles, or nanoscale magnetic particles, have been generating continuous and growing interest since the late 1940s as the investigation of their properties turned out to be very challenging, and each time more promising, from both scientific and technological points of view. In 1949, in a pioneering work, Néel set the pace towards the understanding of the magnetic behavior of nanoparticles, leading to an important development of fundamental theories of magnetism and modeling of magnetic materials, as well as remarkable technological improvements. Magnetic nanoparticles, as compared to bulk materials, enjoy very important novel properties such as enhanced remanence and giant coercivity, and exponentially slow relaxation at low temperature. On the other hand, above a size-dependent temperature they exhibit the phenomenon of *superparamagnetism* which is the result of thermally activated ultrafast crossing of the anisotropy-energy barrier by the magnetic moment of the particle.

Nanomagnetism has provided physicists with a rich and challenging playground for the generalization and application of the concepts and techniques of statistical mechanics at equilibrium and out of equilibrium. In particular, the great efforts devoted to the investigation of nanomagnetism have been rewarded by a noticeable progress in shedding light on our understanding of the finite-size and surface effects on the static and dynamic properties of nano-scaled magnetic systems. Today, active investigation of the fundamental properties of magnetic nanoparticles and the novelties they exhibit, continues with the development of new materials using a variety of modern chemical and physical techniques. Likewise, the characterization and measurements of their properties are becoming more precise than ever thanks to a large choice of cutting-edge new techniques with ever increasing temporal and spatial resolution. This opens up, once again, new perspectives for a better control of the properties of single particles and their more efficient assemblage into more complex structures with ever more promising day-to-day applications such as magnetic recording, magneto-plasmonics, magneto-optics, magneto-calorics, micro- and nano-electronics, magnetic catalysis, and bio-nanomaterials.

Hamid Kachkachi

Chair of the 8th International Conference on Fine Particle Magnetism (ICFPM)

Journal of Physics: Conference Series **521** (2014) 011001 doi:10.1088/1742-6596/521/1/011001

ICFPM 2013 Conference Photo

Journal of Physics: Conference Series **521** (2014) 011002 doi:10.1088/1742-6596/521/1/011002

Proceedings of the 8th

International Conference on Fine Particle Magnetism

24-27 June 2013
Perpignan, France

Guest Editor

Hamid Kachkachi

Perpignan, France

Journal of Physics: Conference Series **521** (2014) 011002 doi:10.1088/1742-6596/521/1/011002

Organized by

The group Nanoscale Spin Systems
Laboratoire PROMES, CNRS UPR 8521
Université de Perpignan Via Domitia
Rambla de la Thermodynamique
Tecnosud - 66100 Perpignan, France
http://www.promes.cnrs.fr/ICFPM2013/home.html

Sponsored by

Région Languedoc-Roussillon
Université de Perpignan Via Domitia

Organizing Committe

H. Kachkachi (Chair)
N. Barros, R. Bastardis, J.-L. Déjardin, J.-M. Martinez, H. Kachkachi, F. Vernay
Secretary staff: S. Payre, A. Leroyer

Conference Chair and Programme Committee

H. Kachkachi (Chair)
Group: Nanoscale Spin Systems (S^2N)
Laboratoire PROMES (CNRS UPR8521)

International Advisory committee Board

Xavier Batlle Gelabert (UB, Barcelona)
Roy Chantrell (York, UK)
Dino Fiorani (ISM – CNR, Italy)
Michael Farle (Duisburg-Essen, Germany)
George Hadjipanayis (Delaware, USA)
Chin-Yih Hong (Chung Hsing, Taiwan)
Hamid Kachkachi (UPVD, France)
Amilcar Labarta (UB, Barcelona)
Hiroaki Mamiya (NIMS, Japan)
Puerto Morales (ICMM, Spain)
Per Nordblad (Uppsala)
Kevin O'Grady (York, UK)
Quentin Pankhurst (UCL, UK)
Dong-Liang Peng (Xiamen, China)
Robert Shull (NIST, USA)

Scientific Comittee

Pascal Andreazza (CRMD, Orléans)
Sylvie Bégin-Colin (IPCMS, Strasbour)
Christine Boeglin (IPCMS, Strasbourg)
Grégoire de Loubens (CEA, Saclay)
Véronique Dupuis (LPMCN, Lyon I)
Oscar Iglesias (IN2, Univ. Barcelona)
Denis Ledue (GPS, Rouen)
Sawako Nakamae (SPEC/CEA-Saclay)
Marc Respaud (INSA, Toulouse)

Journal of Physics: Conference Series **521** (2014) 011002 doi:10.1088/1742-6596/521/1/011002

Sylvie Rousset (MPQ, Paris)
Veronica Salgueiriño (Universidad de Vigo, Spain)
André Thiaville (LPS, Orsay)
Christophe Thirion (Institut Néel, Grenoble)

Peer review statement

All papers published in this volume of *Journal of Physics: Conference Series* have been peer reviewed through processes administered by the proceedings Editors. Reviews were conducted by expert referees to the professional and scientific standards expected of a proceedings journal published by IOP Publishing.

Atomic relaxation effects on magnetism in CoPt nanoalloys

V. Dupuis[1]* and A. Tamion[1]

[1] ILM, UMR 5306 CNRS/Université de Lyon, F-69622 Villeurbanne cedex

Email*: veronique.dupuis@univ-lyon1.fr

Abstract. We describe the magnetism of CoPt clusters upon annealing-driven transition to chemically ordered $L1_0$-like phase. In one hand, X-ray magnetic circular dichroism (XMCD) investigations at each $L_{2,3}$ edge, revealed a significant increase of both spin and orbital of Co and Pt magnetic moments after thermal ordering transition. In the other hand, reversely to the bulk alloys, a limited Magnetic Anisotropy Energy (MAE) enhancement has been obtained from SQUID magnetometry measurements on the same chemically ordered CoPt nanomagnets samples. In this paper, we relate such magnetic behaviour to element-specific dependence of the local atomic relaxations in nanoalloys, described from extended x-ray absorption fine structure (EXAFS) experiments.

1. Introduction

Magnetic nanoalloys attract a lot of attention because they offer the possibility to tune the magnetic moment and the magnetic anisotropy energy (MAE) probably up to the ultimate density storage limit. In particular, an extremely high magnetocrystalline anisotropy is expected from the stacking of pure Co (or Fe) and Pt atomic planes in the (001) direction for CoPt (or FePt) bulk alloys in the chemically ordered $L1_0$ phase. Nevertheless, even if a well chemical order can be observed at nanosize, the achievement of a consequent MAE enhancement without any coalescence after annealing-driven transition, remains so far absent [1, 2, 3, 4]. The MAE determines the low-temperature orientation of the magnetization with respect to the structure of the system. It determines also the energy barrier that fixes the superparamagnetic limit at which the system loses its magnetic memory, a limit that is of crucial importance in non-volatile magnetic memory. The atomic structure and magnetic moment of CoPt and FePt nanoparticles have been experimentally observed to differ from the corresponding bulk materials in a positive way or not, due to small size effects as peculiar symmetry [5], partial chemical ordering [6], surface segregation [7],.. . Numerous theoretical works have been performed to try to explain exotic structures [8, 9, 10, 11] or electronic properties [12,13] observed in nanoalloys by integrating a great number of parameters. In this paper, we relate atomic relaxation effects on magnetism in CoPt nanoalloys.

2. Experimental procedures

2.1. Sample Synthesis

Bi-metallic clusters are preformed in the gas phase thanks to a laser vaporization source working in the Low Energy Clusters Beam Deposition (LECBD) regime [1]. Recently our apparatus is equipped with a quadrupolar electrostatic mass-deviator allowing us to deposit mass-selected clusters in an UHV deposition chamber [5] allowing a gaussian size distribution with a relative standard deviation close to 8 %. We have focused our attention on 3 nm CoPt nanoparticles (NPs) embedded in an inert carbon matrix, in order to preserve and investigate the intrinsic cluster surface properties. We have synthesized size-selected CoPt clusters assemblies, 1%-diluted in volume in C-matrix to quite avoid magnetic interaction among NPs. The idea was to study structure and magnetism of CoPt NPs before and after an optimum annealing of 2 hours at 750 K under high vacuum without any coalescence. First of all, High Resolution TEM-observations and simulations on CoPt NPs revealed that upon annealing a transition occurred up to 2 nm in diameter from as-prepared A1 fcc-structure to a tetragonal chemically ordered $L1_0$ phase with quasi perfect order parameter and truncated octahedron shapes [14].

2.2. EXAFS experiments

The geometrical arrangement has been quantitatively determined from Extended x-ray absorption fine structure (EXAFS) experiments performed in the fluorescence mode on as-prepared and annealed CoPt samples, at both Co-K and Pt-L edge, respectively on the CRG-FAME and ID12 beam lines of the european synchrotron radiation facility (ESRF) at Grenoble [15]. Data were simulated after a 1.2–3.2 Å Fourier-window filtering (distances uncorrected from phase shift). The fits then return the number of nearest neighbors to the absorbing Co or Pt atom for each element N, their average distances R, and their Debye-Waller factors, which are related to the bond-length dispersion. In the Fig.1a, one can see at Co-K edge, a remarkable enhancement in Fourier Transform (FT) intensity upon annealing of the peak located at 2 Å corresponding to the first metallic nearest-neighbors. From FEFF Fit simulations, we clearly show that the Co-Co/Co-Pt coordination ratio changes upon annealing from 1 to 2 as expected for the A1 to $L1_0$ transition with contracted lattice parameters related to the bulk phase [15]. The best simulation has been obtained for a perfect chemically ordered CoPt packing ($L1_0$ like-phase) with a unique Co-Pt first neighbour distance at both edges but surprisingly two different Co-Co and Pt-Pt distances for CoPt clusters [15].

2.3. XMCD experiments

X-ray magnetic circular dichroism (XMCD) signal has been measured on as-prepared and annealed bimetallic nanomagnets CoPt at the Co L-edge on SOLEIL – DEIMOS. Moreover, because induced magnetic moments resulting from 3d-5d proximity effects are expected in non magnetic metal, we also performed (XMCD) measurements at Pt-L edge on ID12 - ESRF beam line on the same samples. In table I, the mean orbital and spin magnetic moments (μ_L and μ_S) per Co and Pt atom, have been determined using the well-known sum rules [16, 17] where the numbers of hole per Co and Pt atoms has been estimated from theoretical band structure calculations (n_h Co = 2.628 and n_h Pt = 2.369) [18].

2.4. SQUID measurements

The magnetic behavior of the cluster-assembled films has been studied using SQUID magnetometry (Superconducting Quantum Interference Device). An accurate analysis has been reached using our recently developed "triple fit" method where the ZFC/FC curves and a room temperature magnetization loop are simultaneously adjusted using a semi-analytical model [19]. By fitting simultaneously the entire magnetic curves, we can reach a reliable and accurate determination of the particle magnetic size distribution and MAE [20]. So, we have found the average magnetic diameter (D_m) is equal to the observed TEM one and remains constant upon annealing. The main difference comes from the MAE evolution between the as-prepared and annealed CoPt sample where the median effective anisotropy constant (K_{eff}) increased and its corresponding dispersion ω_K slightly decreased (see Fig. 1c). This last dispersion decrease is well explained by the appearance of the chemical order at nanosize [21]. But, the MAE value is one order of magnitude smaller that what is expected for the $L1_0$ CoPt bulk.

Journal of Physics: Conference Series **521** (2014) 012001 doi:10.1088/1742-6596/521/1/012001

3. Discussion and conclusion

As a summary, in one hand, we found a significant increase of Co and Pt magnetic moments for the chemically ordered $L1_0$-like clusters structures from X-ray magnetic circular dichroism (XMCD) investigations on CoPt clusters upon annealing. In the other hand, the magnetic anisotropy energy (MAE) of the annealed sample determined from SQUID magnetometry only increased of 50 % compared to the as-prepared sample, i.e. one order of magnitude smaller to what is expected for the $L1_0$ bulk phase. We will try to relate to these specific magnetic results to the element-specific local atomic relaxations in CoPt clusters determined by EXAFS measurements.

First of all, notice that the average spin moment is very sensitive to the numbers of hole per Co and Pt atoms, and to the clusters size but depends weakly on the local atomic distortion. It is the reason why we found an enhancement of both μ_S (Co) and μ_S (Pt) in 3nm-CoPt clusters compared to the one for bulk $L1_0$ phase. Concerning the behavior of the average μ_L (Co) which is generally related to the MAE, we found both that it is more pronounced than the one found for the bulk hcp Co (0.107 μ_B /at.) and also at Co surface film (0.12 μ_B /at.). In fact, the MAE is controlled by the spin-orbit (SO) coupling between occupied and empty electronic states. Since the SO interactions are very sensitive to the position and degeneracy around the Fermi level, significant contributions to the MAE should be expected even if small variations in the local atomic environment are observed. [22]

Table 1. Median magnetic size diameter (D_m), maximum susceptibility (T_{max}), effective anisotropy constant (K_{eff}) and relative dispersion (ω_K) determined from SQUID measurements. Co and Pt spin (μ_S) and orbital (μ_L) magnetic moments from experimental XMCD measurements

	As-Prepared	**Annealed CoPt**
D_m	3.1 nm	3.1 nm
T_{max}	15 K	22 K
K_{eff}	218 KJ/m^3	293 KJ/m^3
ω_K	37 %	28 %
μ_S (Co)	1.67 μ_B */at.*	1.98 μ_B */at.*
μ_L (Co)	0.13 μ_B */at.*	0.20 μ_B */at.*
μ_S (Pt)	0.47 μ_B */at.*	0.52 μ_B */at.*
μ_L (Pt)	0.07 μ_B */at.*	0.10 μ_B */at.*

In order to explain the 3 different distances obtained from EXAFS simulations, we applied to perfect CoPt clusters in the same size range, spin-polarized density-functionnal calculations using the Vienna *ab initio* simulation package (VASP) [15]. In contrast to the bulk structure, the Co and Pt layers appear to relax independently from each other. In particular at cluster surface, the Co atoms show a clear in-plane tetramerization in the uppermost (001) Co layer, which does not match the underlying Pt layer. So, despite the strong CoPt cluster hybridization, we also found 3 different calculated relaxed mean distances for Co-Co, Co-Pt and Pt-Pt with gaussian shape distributions, in very good agreement with EXAFS experimental values. In conclusion, while in the bulk $L1_0$ phases, an equal Co-Co and Pt-Pt Dirac distance distribution is expected, a strong Co-Co distance (d_{CoCo}) dispersion is predicted and experimentally observed for relaxed clusters even in the chemically ordered phase (see Fig. 1.b).

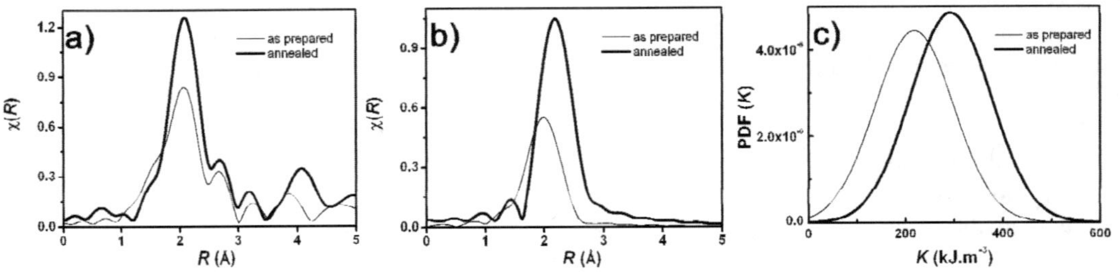

Figure 1. Experimental Fourier Transform $\chi(R)$ at Co-K edge for first metallic nearest-neighbors (a), Co-Co first-neighbor contribution to the $\chi(R)$ from its best FEFF simulation (b), and magnetic anisotropy constant *Keff* dispersion (c) for mass-selected CoPt sample before and after annealing.

In a previous paper [23], we have shown that the MAE dispersion in pure Co clusters essentially comes from the effect of additional facets and is relatively small. But in nanoalloys, the size and shape dispersion is not the unique source of MAE dispersion. Indeed, since the anisotropy enhancement in as-prepared CoPt compared to pure Co clusters is due to the presence of Pt atoms, the dispersion of the magnetocrystalline anisotropy (which depends on the neighborhood of each Co atom) increases with the number of possible chemical arrangements. It is the reason why the MAE of chemically disordered CoPt particles is quite large even if mass-selected clusters have a small size dispersion (8%) and a highly symmetrical shape (regular truncated octahedron)[2]. As long as a well-defined and high enough degree of chemical order can be reached, the multiplicity of atomic configurations is strongly reduced and the MAE dispersion decreases while its median value increases [14, 21]. But at the same time, the relaxation dispersion increases due to the stress between alternating Co and Pt planes (see figure 2 in [24]). In ref [22], by investigating different relaxations, it is shown using semi-empirical model, that the existence of nonuniform pattern of interatomic distances in 3d transition metal (TM) clusters can cause appreciable changes in the magnitude of the MAE. In particular, it is shown that negative and positive values for local MAE at different clusters sites can lead to an almost vanishing of the global MAE in magnetic clusters due to the cancellation of all these local contributions.

So we claim that we have obtained new experimental insights on the correlation between magnetic properties and short- or long-range chemical order in nanoalloys. We evidenced that specific finite size and more specifically strong relaxation effects are certainly related to the low MAE results obtained even on nearly perfect chemical ordered bimetallic nanoparticles assemblies.

To go further, non-collinear calculations, including the spin-orbit coupling in order to quantitatively access to the MAE values are to be done for such relaxed nanoalloys (Ab-initio+ tight-binding calculations).

Acknowledgment

The authors are grateful to A. Ramos, F. Wilhelm and E. Otero for their stimulating discussions and their help during experiments on CRG-FAME and ID12 beam lines at ESRF and on DEIMOS at SOLEIL. Support is acknowledged from both GDR CNRS 3182 and COST-STSM-MP0903-7318 on Nanoalloys.

References

1 F. Tournus, A. Tamion, N. Blanc, A. Hannour, L. Bardotti, B. Prevel, P. Ohresser, E. Bonet, T. Epicier, and V. Dupuis, Phys. Rev. B **77**, 144411 (2008).

2 F. Tournus, N. Blanc, A. Tamion, M. Hillenkamp and V. Dupuis, Phys. Rev. B **81**, 220405(R) (2010).

3 C.-b. Rong, N. Poudyal, G. S. Chaubey, V. Nandwana, R. Skomski, Y. Q. Wu, M. J. Kramer, J. P. Liu, J. Appl. Phys. **102**, 043913 (2007).

[4] X. Sun, Z. Y. Jia, Y. H. Huang, J. W. Harrell, D. E. Nikles, K. Sun, and L. M. Wang, J. Appl. Phys. **95**, 6747 (2004).
[5] F. Tournus, N. Blanc, A. Tamion, M. Hillenkamp, V. Dupuis, *J. Magn. Magn. Mater.* **323**, 1868 (2011) ; F. Tournus, K. Sato, T. J. Konno, T. Epicier, V. Dupuis, Phys. Rev. Letters **110**, 055501 (2013).
[6] N. Jaouen, D. Babonneau, J. M. Tonnerre, D. Carbone, F. Wilhelm, A. Rogalev, T. K. Johal, and G. van der Laan, Phys. Rev. B **76**, 104421 (2007).
[7] Antoniak, J. Lindner, M. Spasova, D. Sudfeld, M. Acet, M. Farle, K. Fauth, U. Wiedwald, H.-G. Boyen, P. Ziemann, F. Wilhelm, A. Rogalev, and S. Sun, Phys. Rev. Lett. **97**, 117201 (2006).
[8] G. Rossi, R. Ferrando, and C. Mottet, Faraday Discuss. Chem. Soc. **138**, 193 (2008).
[9] M. E. Gruner, G. Rollmann, P. Entel, and M. Farle, Phys. Rev. Lett. **100**, 087203 (2008).
[10] P. Entel, M. E. Gruner, G. Rollmann, A. Hucht, S. Sahoo,, T. Zayak, H. C. Herper, and A. Dannenberg, Philos. Mag. **88**, 2725 (2008).
[11] P. Andreazza, C. Mottet, C. Andreazza-Vignolle, J. Penuelas, H. C. N. Tolentino, M. De Santis, R. Felici, and N. Bouet, Phys. Rev. B **82**, 155453 (2010).
[12] C. Barreteau and D. Spanjaard, J. Phys.: Condens. Matter **24** 406004 (2012).
[13] R. Cuadrado and R. W. Chantrell, Phys. Rev. B **86**, 224415 (2012).
[14] N. Blanc, F. Tournus, V. Dupuis, T. Epicier, Phys. Rev. B **83**, 092403 (2011) as Brief Report.
[15] Blanc N., Diaz-Sanchez L. E., Ramos A. Y., Tournus F., Tolentino H. C. N., De Santis M., Proux O., Tamion A., Tuaillon-Combes J., Bardotti L., Boisron O., Pastor G. M., Dupuis V., *Phys. Rev. B*, **87**, 155412 (2013) ; V. Dupuis, N. Blanc, L-E. Diaz-Sanchez, A. Hillion, A. Tamion, F. Tournus, G-M. Pastor, A. Rogalev and F. Wilhelm, Eur. Phys. J. D **67**: 25 (2013).
[16] B. T. Thole, P. Carra, F. Sette, and G. van der Laan, Phys. Rev. Lett. **68**, 1943 (1992).
[17] P. Carra, B. T. Thole, M. Altarelli, and X. Wang, Phys. Rev. Lett. **70**, 694 (1993).
[18] I. Galanakis, M. Alouani and H. Dreysse: J. Magn. Magn. Mater. **242–245** 27(2002).
[19] A. Tamion, M. Hillenkamp, F. Tournus, E. Bonet, V. Dupuis, Appl. Phys. Lett. **95** 062503 (2009).
[20] V. Dupuis, N. Blanc, F. Tournus, A. Tamion, J. Tuaillon-Combes, L. Bardotti, O. Boisron, IEEE Trans. Magn. **47**, 3358 (2011).
[21] F. Tournus, S. Rohart and V. Dupuis**,** IEEE Transactions on Magnetics **44** 3201 (2008).
[22] R. Guirado-Lopez, Phys. Rev. B **63**, 174420 (2001).
[23] M. Jamet, W. Wernsdorfer, C. Thirion, D. Mailly, V. Dupuis, P. Mélinon, and A. Pérez, Phys. Rev. Lett. **86**, 4676 (2001).
[24] V. Dupuis, N. Blanc, L-E. Diaz-Sanchez, A. Hillion, A. Tamion, F. Tournus, G-M. Pastor, Eur. Phys. J. B **86**: 83 (2013).

Concentric $MnFe_2O_4$-rich core/Cr_2O_3-rich shell nanoparticles

S. Liébana-Viñas,[1] R. Otero-Lorenzo,[1] B. Rivas-Murias,[2] V. Salgueiriño[1,1]
[1] Departamento de Física Aplicada, Universidade de Vigo, 36310, Vigo (Spain)
[2] Centro de Investigación en Química Biológica y Materiales Moleculares (CIQUS), Universidad de Santiago de Compostela, 15782-Santiago de Compostela, Spain

E-mail: vsalgue@uvigo.es

Abstract. Concentric $MnFe_2O_4$-rich core/Cr_2O_3-rich shell nanoparticles were obtained considering a postsynthetic modification of preformed $MnFe_2O_4$@$Cr(OH)_3$ units. The complex nanomaterials display an exchange bias (EB) effect likely due to the interface intermixing and possible cluster spin-glass phase and the presence in the same units of ferrimagnetic and antiferromagnetic-ferroelectric oxides. This particular case of transition metal oxides (TMO) heterostructures underlines the potential of chemical control in their final magnetic properties.

1. Introduction

A material is generally considered functional if it possesses a physical property that is usable in applications. Therefore considerable amount of research currently concentrates on multifunctional materials in which several physical properties could potentially be used simultaneously.[1] Multiferroics for example, represent a promising option because they simultaneously possess several ferroic orders such as ferromagnetism and ferroelectricity.[2] Such combination of functionalities can be exhibited by TMO in heterostructures,[3] bearing in mind the semilocalized character of their d electrons.[4] The correlation between chemical composition and structure in nanoscale systems particularly, considering ion and vacancy diffusion lengths, becomes a dynamical variable to control the final properties of nanocomposites. For example, changes in cationic valence state associated with changes in oxygen vacancy concentration profoundly affect magnetic, electronic and transport properties.[5] Accordingly, developments in synthesis to exploit and control the different mobility of cations and anions to engineer gradients in static and dynamic chemistry and carrier concentrations are required.[6] Herein we report concentric structured nanocomposites based on ferrimagnetic (FiM) $MnFe_2O_4$ and antiferromagnetic (AFM) and ferroelectric (FE) Cr_2O_3. The synthesis includes driving $Cr(OH)_3$ on the surface of clusters of $MnFe_2O_4$ nanoparticles, and a post-synthetic thermal treatment that renders the nanostructures Mn ferrite- and chromium oxide-rich at the core and shell, respectively, and offering concentric phases of Mn ferrite, iron and chromium oxides that determine the final magnetic behavior.

2. Experimental

$MnFe_2O_4$ clusters synthesis was carried out according to previous work.[7] $FeCl_3$ (5 mmol (Fluka) and $MnCl_2$ (2.5 mmol) were dissolved in ethylene glycol (40 mL (Aldrich)), followed by addition of NaAc

[1] To whom any correspondence should be addressed.

(3.6 g (Aldrich)) and polyethylene glycol (1.0 g (Aldrich)). The mixture was stirred for 30 min and sealed in a teflonlined stainless-steel autoclave, heated to and maintained at 185 °C for 8 h, and then allowed to cool down. The black products were washed and dispersed in aqueous solution. As a second step, 1 mL of $Cr(NO_3)_3$ (0.1 M (Aldrich)) and 4 mL of NaOH (0.075 M (Aldrich)) were added dropwise under mechanical stirring to 6 mL of $MnFe_2O_4$ clusters (1.55 mg/mL), keeping T at 25 °C. After 10 min, the composites formed were centrifuged, dried and annealed at 650°C during 2h in air.

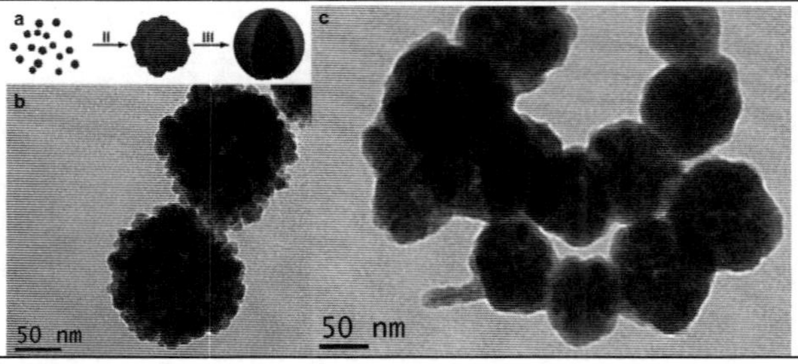

Figure 1. Scheme (a) of the synthesis and TEM images of the clusters (b) and composites (c).

3. Results and Discussion

$MnFe_2O_4$ clusters were synthesized modifying the process reported by Li *et al.*[7] to have a suitable diameter (108 ± 10 nm) for the process herein reported and schematically shown in fig. 1a. These cores (see fig. 1b) once in aqueous solution permit a controlled deposition of $Cr(OH)_3$ in basic medium onto their surface,[8] such that the hydroxide precipitates and evolves into a thin layer surrounding them to form core-shell nanostructures. Once the core-shell structures were formed, they were centrifuged, washed and annealed at 650°C in air to promote the formation of the outer Cr_2O_3 shell. The TEM analysis (fig. 1c) reveals the core-shell structured attained, with a clear variation in the contrast up to the surface of the nanoparticles and likely with the chromium oxide (< 8 nm thick) surrounding each Mn ferrite-based multiple core. The XRD pattern of the sample (fig. 2a) corresponds to a mixture of several phases including the expected chromium oxide (Cr_2O_3, (JPCDS-ICDD card 38-1479)) as the main component (indexed) and Mn ferrite ($MnFe_2O_4$, (JPCDS-ICDD card 10-0319)) with the lower percentage in the sample. The thermal treatment promotes an intermixing process favoring the formation of hematite (Fe_2O_3, (JPCDS-ICDD card 24-0072)) and chromium-iron oxide ($Cr_{2-x}Fe_xO_3$). Indeed, kinetic factors become key in determining the feasibility of an ion exchange reaction, especially at the nanoscale at which limits on long-range solid-state ion diffusion are significantly relaxed by the large surface-to-volume ratios.[9] These results were confirmed by Raman spectroscopy (fig. 2b) which also demonstrated the intermixing of phases in the initial cores.

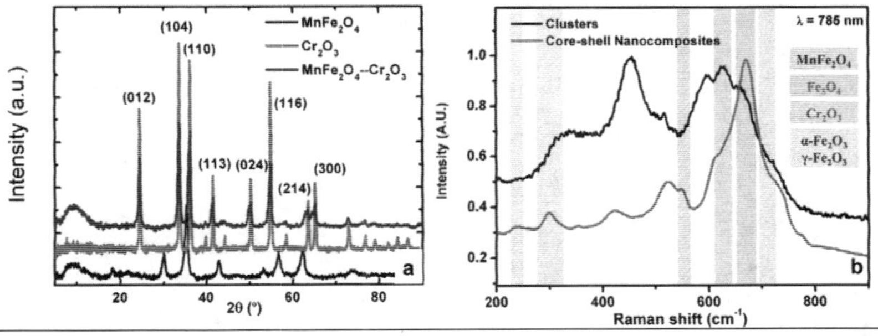

Figure 2. XRD pattern of the sample (a) and Raman spectra of the clusters before and after coated and comparison with main phases spectra (b).

The hysteresis loops registered at 10 K (fig. 3a) show an important decrease in the net value of the magnetization per mass of sample as shifting from $MnFe_2O_4$ (calcinated) clusters to the nanocomposites. A characteristic superparamagnetic behavior with low values of H_C, which decrease notably as the temperature increase, is observed in both cases. This feature underlines the presence of small magnetic units interacting in the samples, even after the annealing process. Fig. 3b includes the temperature dependence of magnetization (ZFC-FC (100 Oe)) and in both cases the curves split around T=300 K, which can also be associated with a strongly interacting system.

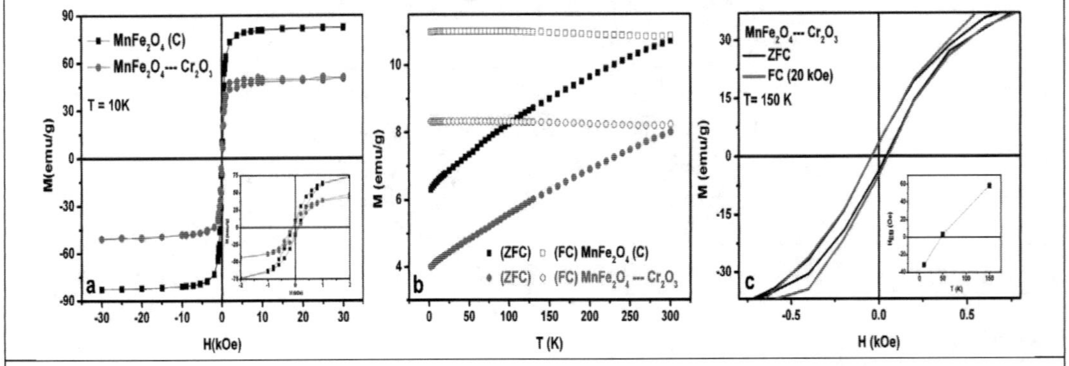

Figure 3. MvsH and MvsT of $MnFe_2O_4$ (calcinated) cores and $MnFe_2O_4$ - Cr_2O_3 nanocomposites.

The difference in the magnetic behavior of both samples can be related to the coupling between the FiM ($MnFe_2O_4$ and $Cr_{2-x}Fe_xO_3$[10]) and the AFM (hematite (T_N = 956 K) and Cr_2O_3 (T_N = 310 K)) phases taking place in the heterostructures. The hysteresis loops (fig. 3c) at 150 K (ZFC and FC (20 kOe) reflect a subtle increase in H_C, around 100 Oe, and shift in the second case (likely due to averaging effects as having a 3D connectivity pattern between the FiM-AFM interfaces), which are generally attributed to an EB effect.[11] Though the interfacial interaction between the FiM-AFM phases may be blurred by the intermixing, it appears to be the main source of EB. The EB field ($H_{EB} = (H_{C+}+H_{C-})/2$) varies with temperature It also exhibits a temperature-driven sign reversal related to a change in the magnetic order at the interface and likely due to a spin-glass interaction at low temperatures, favored by the polycrystalline nature of the strong interacting units in the sample (see inset in fig. 3c).[12,13]

4. Conclusion

Concentric structured nanoparticles of FiM $MnFe_2O_4$ and AFM and FE Cr_2O_3 with a mixed interface based on hematite and chromium iron oxide were prepared. These complex nanomaterials display an exchange bias effect likely due to the mentioned interface and possible cluster spin-glass phase.

5. References

[1] Kreisel J and Kenzelmann M, 2009 *Europhysics News* 40 17
[2] Khomskii D 2009 *Physics* 2 20, Bibes M and Barthélémy A 2008 *Nature Mater.* 7 425
[3] Zubko P, Gariglio S, et al. 2011 *Annu. Rev. Condens. Matter Phys.* 2 141
[4] Goodenough J B 1971 *Metallic Oxides. Progress in Solid State Chem.* (Pergamon Press)
[5] Kalinin S and Spaldin N A 2013 *Science* 341 858
[6] Palkut M, Wiik K and Grande T 2007 *J. Phys. Chem. C* 111 813
[7] Deng H, Li X, Peng Q, Wang X, Chen J, Li Y, 2005 *Angew. Chem.* 44 2782
[8] Bañobre-López M, Vázquez C, Rivas J and Quintela M A L 2003 *Nanotechnology* 14 318
[9] Beberwyck B J, Surendranth Y and Alivisatos A P 2013 *J. Phys. Chem. C* 117 19759
[10] Sadat Nabi H and Pentcheva R 2011 *Phys Rev B* 83 214424

[11] Meiklejohn W H and Bean C P 1956 *Phys. Rev* 102 1413, Nogués J and Schuller I K 1999 *J. Magn. Magn. Mater.* 192 203

[12] Salgueiriño-Maceira V, Correa-Duarte M A, Bañobre-López et al. 2008 *Adv. Func. Mat* 18 616

[13] Rizwan Ali S, Bilal Janjua M, et al. 2010 *Phys. Rev. B* 82 020402

Acknowledgments

B. R.-M. acknowledges her post-doc position from ERC (StG-259082, 2DTHERMS). V. S. acknowledges the financial support from the *Ramón y Cajal* Program and project 10PXIB312260PR (M. de Economía and Xunta de Galicia (Regional Government, Spain)).

Investigating individual $Fe_{50}Co_{50}$ alloy nanoparticles using X-ray photo-emission electron microscopy

A. Kleibert[1], A. Balan[1], A. Fraile Rodríguez[2], and F. Nolting[1]

[1]Swiss Light Source, Paul Scherrer Institut, Villigen PSI, CH-5232 Switzerland
[2]Departament de Física Fonamental and Institut de Nanociència i Nanotecnologia (IN2UB), Universitat de Barcelona, E-08028 Barcelona, Spain.

Email: armin.kleibert@psi.ch

Abstract. *In situ* X-ray photo-emission electron microscopy (PEEM) together with X-ray circular magnetic dichroism (XMCD) and complementary *ex situ* scanning electron microscopy (SEM) is used to investigate the properties of individual $Fe_{50}Co_{50}$ alloy nanoparticles in a mass-filtered particle ensemble with a mean size of 12 nm. The data reveal a uniform chemical composition of the particles, but a wide distribution of magnetic properties resulting in a co-existence of superparamagnetic (SPM) and ferromagnetically blocked particles. The ferromagnetic (FM) particles suggest enhanced magnetic energy barriers when compared to the magneto-crystalline anisotropy energy (MCA) of bulk $Fe_{50}Co_{50}$.

I. Introduction

Magnetic nanoparticles have attracted recently considerable research interest due to their current and prospective applications in spintronics, drug delivery, cancer therapy, multifunctional materials, and ultrahigh density data storage [1,2,3]. The latter require nanomagnets with magnetic energy barriers E_m which prevent thermally driven switching of the magnetization at room temperature together with a large saturation magnetization to promote reading and writing information at smaller bit sizes [4]. CoPt or FePt alloys in their chemically ordered $L1_0$ phase provide the required properties. In particular, they possess a high MCA providing a respectively high E_m, but they are difficult to prepare at the nanoscale [5]. More recently, the use of FeCo alloys was proposed, which have only small MCA in the bcc bulk phase, but can develop an MCA similar to that of FePt when the lattice is tetragonally distorted [4]. Such distortions and an increased MCA have indeed been found in FeCo thin films [6].

Recent investigations on FeCo nanoparticles revealed a high saturation magnetization and indication for enhanced E_m when compared to the bulk MCA of bcc FeCo [7,8]. However, the data were obtained from ensemble measurements and thus, gave no direct insight in the distribution of properties within these ensembles. A significant scattering of properties can for instance arise from size and shape variations, inter-particle interactions, as well as the contact with the surrounding medium and local surface effects of the individual particles [9]. In case of alloys, the situation becomes even more complicated due possible particle-to-particle variations of the composition and chemical order [5]. Recently, it was demonstrated that X-ray PEEM is suitable to probe the chemical composition and magnetic properties of individual nanoparticles [9,10,11]. In this contribution we use this technique to investigate the distribution of composition and magnetic properties of single FeCo nanoparticles in a mass-filtered particle ensemble.

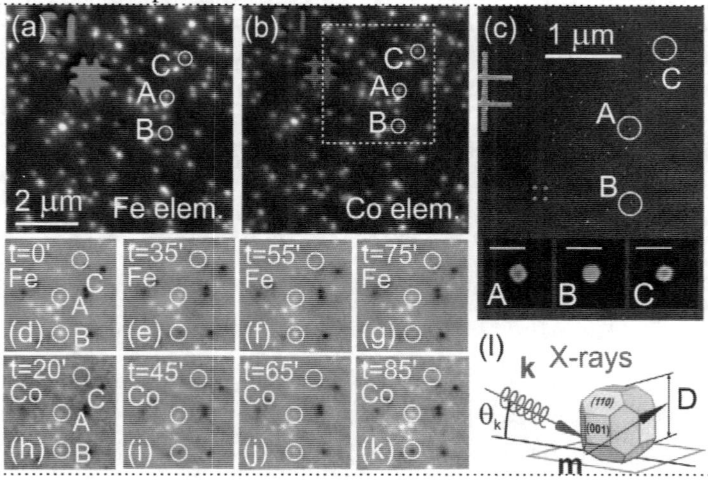

Figure 1. (a),(b) Elemental contrast images recorded at (a) the Fe and (b) the Co L_3 edge. (c) SEM image of the area denoted with the dashed box in (b). The insets show high resolution SEM images of three particles denoted as A, B, and C in (a)-(k). The scale bar in the inset indicates a length of 50 nm. (d)-(g) Magnetic contrast images recorded in the denoted area in (b) as a function of time *t* (given in minutes) at the Fe L_3 edge. (h)-(k) similar to (d)-(g), but taken at the Co L_3 edge. The particles A, B, C from (b) and (c) are indicated. (l) Sketch of the experimental geometry and the expected equilibrium shape of bcc FeCo nanoparticles.

2. Experimental Details

The nanoparticles are generated in the gas phase by means of an arc cluster ion source using a $Fe_{50}Co_{50}$ alloy target (Vacoflux 50, Vacuumschmelze GmbH) [12]. Mass-filtered particles with a mean size D = 12 nm and size distribution of $\Delta D/D$ = 0.15 [12] are then deposited onto Si wafers with a native oxide layer. A gold mesh is used to measure the particle flux during deposition and to ensure a sufficiently low particle density of a few particles per µm^2 which is required to avoid inter-particle interactions and to enable single particle investigations in our samples. Prior to the particle deposition, the substrates are cleaned by Ar sputtering (20 min, 0.5kV, $5\cdot10^{-5}$ mbar). The elemental composition and the magnetic properties of the particles are studied *in situ* using X-ray PEEM at a base pressure below $5\cdot10^{-10}$ mbar [13]. In the microscope the samples are illuminated at a grazing angle θ_k = 16° with respect to the surface [Fig. 1(f)]. Details on the acquisition of elemental and magnetic contrast maps can be found elsewhere [10,11,14]. All experiments are carried out at room temperature. The particle morphology is studied subsequently to the PEEM experiments by *ex situ* SEM. Lithographic Au/Cr markers on the substrates enable us to identify the same particles in SEM and PEEM.

3. Results and discussion

Elemental contrast maps recorded with X-ray PEEM reveal bright spots at the same sites in images obtained with the photon energy tuned to the Fe and Co L_3 edges, cf. Figs. 1(a) and (b), respectively. Similar to earlier work on size-dependent properties of individual Fe nanoparticles [11,14], most of the bright spots can be assigned to individual nanoparticles with a well-defined, compact shape, as demonstrated by SEM, cf. Fig. 1(c) and the insets. Clusters of particles and agglomerates are excluded from the analysis. Note that the apparent particles size in SEM is affected (increased) due to the formation of an oxide shell with a thickness of 2-3 nm surrounding the $Fe_{50}Co_{50}$ cores upon sample transfer in ambient air in addition to the limited spatial resolution of the SEM which is about 1 – 2 nm. A detailed correlation of the SEM images with the PEEM elemental contrast maps reveals that every nanoparticle indeed consists of both elements Fe and Co, respectively. Thus, a decomposition of the alloy target material into separate Fe and Co nanoparticles can be excluded.

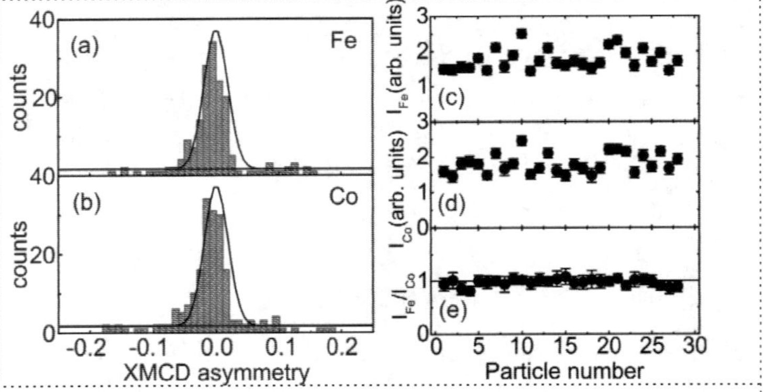

Figure 2. (a),(b) Element-specific histograms of the XMCD asymmetry of individual FeCo nanoparticles. The black and blue lines serve as a guide to the eye. (c),(d) Element-specific intensities I_{Fe} and I_{Co} determined from the elemental contrast maps in Figs. 1(a),(b) for a number of particles.(e) Ratio I_{Fe}/I_{Co} of the same particles.

Examples of element-specific magnetic contrast images, recorded as a function of time in the area denoted in Fig.1(b), are shown in Figs. 1(d)-(k). The magnetic contrast of the particles ranges from black to white through a variety of grey values for both Fe and Co constituents. The magnetic contrast is proportional to the projection of the magnetization m onto the X-ray propagation vector k [11]. Thus, the variety of grey tones reflects a random orientation of the magnetization of the particles due the stochastic deposition process [15]. Some of these particles show stable magnetic contrast over the total observation time which was 9 hours, such as particle "A" in Fig. 1. The presence of such FM particles is not expected when considering FeCo nanoparticles with bulk-like properties and a highly symmetric equilibrium shape, as schematically shown in Fig.1(l) [12]. Assuming a typical attempt frequency of $1\cdot 10^9 s^{-1}$ and using the cubic bulk MCA of FeCo ($K_1 \sim 1\cdot 10^4 J/m^3$) in the Arrhenius law for magnetic relaxation, we find that at room temperature only particles with a size larger than about 45 nm should show stable magnetic contrast over the full observation period. This suggests, that the observed FM particles demonstrate significantly modified properties when compared to the above model assumptions.

Closer inspection of the data reveals also the presence of SPM behaviour in a number of particles, which in some cases can be directly observed as changes of sign and magnitude of the magnetic contrast of such particles in time, cf. particle "B" in Figs. 1(d)-(k). The respective fluctuations occur with a rate of a few switching events per hour and thus indicate the onset of SPM behaviour. SPM particles which fluctuate at a much higher rate show no magnetic contrast in all images due to the vanishing average magnetization. Such a case might be presented by particle "C" in Figs. 1(d)-(i). The actual SPM proportion on the sample can be seen, when plotting histograms of the XMCD asymmetry (cf. Ref. [11]) as shown in Figs. 2(a) and (b), where the peak around zero XMCD (indicated by the black line) reflects the SPM particles (about 60% of the ensemble). The flat part (indicated by the blue line) can be assigned to the FM particles with random orientation of m. Notably the histograms for Fe and Co are similar, as expected from an alloy with well coupled Fe and Co moments.

The distribution of chemical compositions can be studied when analyzing the particle intensities in elemental Fe and Co contrast maps [Figs. 1(a) and (b)]. In Figs. 2(c) and (d) we display the respective intensities I_{Fe} and I_{Co} for a number of particles. The data reveal typical particle-to-particle variations of the intensity as discussed in Ref. [14]. However, the ratio I_{Fe}/I_{Co} is almost constant, which indicates a similar amount of Fe and Co in the probed volume of each particle. Note that the ratio I_{Fe}/I_{Co} does not directly reflect the atomic ratio of Fe and Co, since the absorption cross sections of Fe, Co, and the Si substrate at the Fe and Co L_3 edge energies are different.

4. Conclusion

We have used X-ray PEEM to study the composition and magnetic properties of individual FeCo alloy nanoparticles. The data revealed a wide distribution of magnetic energy barriers which result in a coexistence of superparamagnetic and ferromagnetically blocked nanoparticles in a mass-filtered ensemble of particles with a mean size of 12 nm. In contrast, the atomic ratio of Fe and Co in the probed volume of the particles was found to be uniform in the sample. However, local inhomogeneities in the individual nanoparticles are not excluded and might be responsible for the wide distribution of magnetic properties [16]. The ferromagnetic nanoparticles indicate increased magnetic energy barriers when compared to estimates based on the magneto-crystalline anisotropy of bulk FeCo. Further studies are needed to reveal whether the enhanced magnetic energy barriers are indeed due to the proposed tetragonal lattice distortions or due to other effects.

Acknowledgements

Part of this work was performed at the Surface/Interface:Microscopy (SIM) beam line of the Swiss Light Source, Paul Scherrer Institut, Villigen, Switzerland. This project was funded by the Swiss Nanoscience Institute, University of Basel. AFR acknowledges support from the MICIIN "Ramón y Cajal" Programme.

References

[1] Bansmann J, Baker S H, Binns C, Blackman J A, Bucher J P, Dorantes-Dávila J, Dupuis V, Favre L, Kechrakos D, Kleibert A, Meiwes-Broer K H, Pastor G M, Perez A, Toulemonde O, Trohidou K N, Tuaillon J and Xie Y 2005 *Surf. Sci. Rep.* **56** 189

[2] Rusponi S, Cren T, Weiss N, Epple M, Buluschek P, Claude L and Brune H 2003 *Nat. Mater.* **2** 546

[3] Antoniak C, Gruner M E, Spasova M, Trunova A V, Römer F M, Warland A, Krumme B, Fauth K, Sun S, Entel P, Farle M and Wende H 2011 *Nat. Comm.* **2** 528

[4] Burkert T, Nordström L, Eriksson O and Heinonen O 2004 *Phys. Rev. Lett.* **93** 027203

[5] Tournus F, Sato K, Epicier T, Konno T J and Dupuis V 2013 *Phys. Rev. Lett.* **110** 055501

[6] Andersson G, Burkert T, Warnicke P, Björck M, Sanyal B, Chacon C, Zlotea C, Nordström L, Nordblad P and Eriksson O 2006 *Phys. Rev. Lett.* **96** 037205

[7] Getzlaff M, Bansmann J, Bulut F, Gebhardt R K, Kleibert A and Meiwes-Broer K H 2006 *Appl. Phys. A* **82** 95

[8] Getzlaff M, Kleibert A, Methling R, Bansmann J and Meiwes-Broer K H 2004 *Surf. Sci.* **566-568** 332

[9] Kronast F, Friedenberger N, Ollefs K, Gliga S, Tati-Bismaths L, Thies R, Ney A, Weber R, Hassel C, Römer F M, Trunova A V, Wirtz C, Hertel R, Dürr H A and Farle M 2011 *Nano Lett.* **11** 1710

[10] Fraile Rodríguez A, Nolting F, Bansmann J, Kleibert A and Heyderman L J 2007 *J. Magn. Magn. Mater.* **316** 426

[11] Fraile Rodríguez A, Kleibert A, Bansmann J, Voitkans A, Heyderman L J and Nolting F 2010 *Phys. Rev. Lett.* **104** 127201

[12] Kleibert A, Passig J, Meiwes-Broer K H, Getzlaff M and Bansmann J 2007 *J. Appl. Phys.* **101** 114318

[13] Le Guyader L, Kleibert A, Fraile Rodríguez A, El Moussaoui S, Balan A, Buzzi M, Raabe J and Nolting F 2012 *J. Elec. Spectr. Rel. Phen.* **185** 371

[14] Fraile Rodríguez A, Kleibert A, Bansmann J, and Nolting F 2010 *J. Phys. D: Appl. Phys.* **43** 474006

[15] Kleibert A, Voitkans A and Meiwes-Broer K H 2010 *Phys. Rev. B* **81** 073412

[16] Antoniak C, Lindner J, Spasova M, Sudfeld D, Acet M, Farle M, Fauth K, Wiedwald U, Boyen H G, Ziemann P, Wilhelm F, Rogalev A, and Sun S, 2006 *Phys. Rev. Lett.* **97** 117201

Particle size effects on the magnetic behaviour of 5 to 11 nm Fe_3O_4 nanoparticles coated with oleic acid

K Chesnel[1], M Trevino[1], Y Cai[1], J M Hancock[2], S J Smith[2] and R G Harrison[2]
[1] Department of Physics and Astronomy, BYU, Provo, Utah, 84602, USA
[1] Department of Chemistry and Biochemistry, BYU, Provo, Utah, 84602, USA

E-mail: kchesnel@byu.edu

Abstract. The magnetic behaviour of 5 to 11 nm magnetite (Fe_3O_4) nanoparticles (NPs) was measured at various temperatures (*T*s) from 20 to 400 K. The particles were fabricated via thermal decomposition of an iron precursor, and involved a coating of the particles with oleic acid. The particle size distribution was analysed by XRD measurements and TEM imaging. Magnetization loops, measured at various *T*s, indicate a superparamagnetic behaviour at high *T* and the occurrence of hysteresis at low *T*, with a stronger coercivity for the larger particles. Zero-Field-Cooling (ZFC) and Field Cooling (FC) curves indicate a superparamagnetic behaviour, with a blocking temperature varying significantly with the particle size. Namely, the peak temperature, T_{max}, increases from 30 K to 170 K when the particle size increases from 5 nm to 11nm. Magnetic couplings between particles appear stronger for larger particles.

Magnetic NPs are key materials for a variety of applications in nanotechnologies and biomedicine.[1] Magnetite (Fe_3O_4) NPs are excellent candidates for medical applications because of their non-toxicity and ability to be highly functionalized.[2] The magnetic properties of bulk magnetite Fe_3O_4, which orders ferrimagnetically below T_c ~850K, have been widely studied. These bulk properties are however altered when magnetite is nanometric in size. Fe_3O_4 NP assemblies exhibit superparamagnetic behaviors, where the NPs behave as single macrospins and collectively reach a frozen state at low *T*. One important question is how this magnetic behaviour is modified by NP size. In this paper, we report magnetic measurements of Fe_3O_4 NPs, ranging from 5 to 11 nm in size. Results are discussed and compared to other measurements of Fe_3O_4 NPs reported elsewhere.

We synthesized our Fe_3O_4 NPs by thermal decomposition of an iron precursor and coated the NPs with oleic acid. Different procedures and reaction temperatures were used to achieve different particle sizes. The 5 nm NPs (sample A), were fabricated by heating iron(III) oleate in oleic acid and octadecene to 300 °C for 30 min, and cooled to room T.[3,4] The 8 nm NPs (sample B), were fabricated from Fe(III) acetylacetonate, mixed with hexadecane, octadecene, oleic acid and oleyamine, and heated to 200 °C for 30 minutes, then heated under nitrogen at 290 °C for another 30 minutes, and cooled to room T.[5] The 11 nm NPs (sample C), were prepared following a similar procedure to sample A, but heated at 320 °C for 30 min. All the NPs were precipitated with ethanol and decanted.

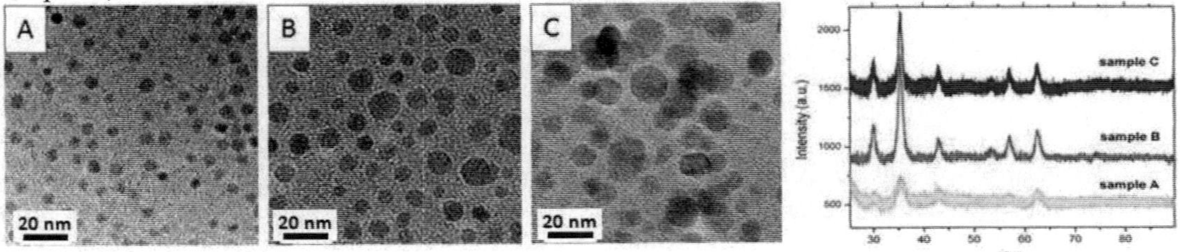

Figure 1: TEM images and XRD patterns of the Fe_3O_4 nanoparticles (A) 5 nm (B) 8 nm (C) 11 nm.

Journal of Physics: Conference Series **521** (2014) 012004 doi:10.1088/1742-6596/521/1/012004

Transmission electron microscopy (TEM) images of the three NP samples are shown in Figure 1. A statistical analysis of the NP sizes yields an average particle diameter of 5.3 ± 0.7 nm, 8.1 ± 1.7 nm, and 11.3 ± 2.5 nm for samples A, B and C, respectively. The X-ray diffraction (XRD) patterns all exhibit a cubic *Fd3m* structure, consistent with Fe_3O_4 magnetite, with no sign of hematite (γ-Fe_2O_3). Average crystallite sizes were estimated from the peak widths: 5.8 ±1.7nm, 8.5 ±2.9 nm and 11.0 ± 4.6 nm for samples A, B and C, respectively, which are in agreement with the TEM results, suggesting monocrystalline NPs. Larger size distributions for sample B and C may have been due to temperature fluctuations during synthesis. Figure 2a shows VSM magnetization curves for the 8 nm NPs, measured at 400, 300, 80 and 20 K. The curves indicate a superparamagnetic behaviour at 400K, well fitted by the Langevin model. The curves at lower T gradually deviate from that behaviour. When T decreases, the susceptibility at $H = 0$ increases and the NPs magnetically align more easily. Also, significant hysteresis occurs at 20 K, indicating a blocked state where the NPs show a ferromagnetic behaviour.

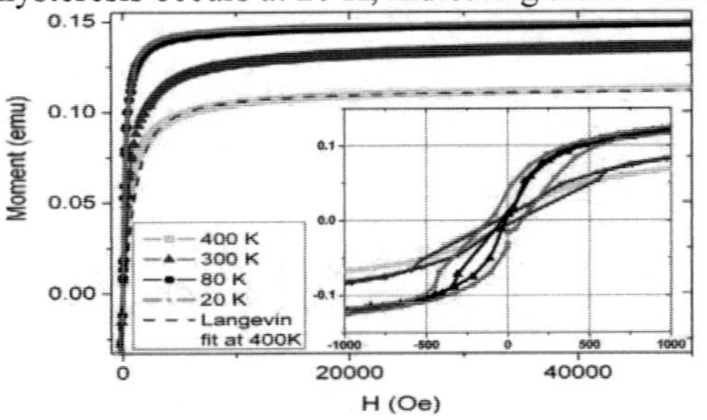

Figure 2. Magnetization curves of sample B (8 nm) at 400, 300, 80 and 20 K, with Langevin fit at 400K. Inset: close-up on the hysteresis at $H = 0$.

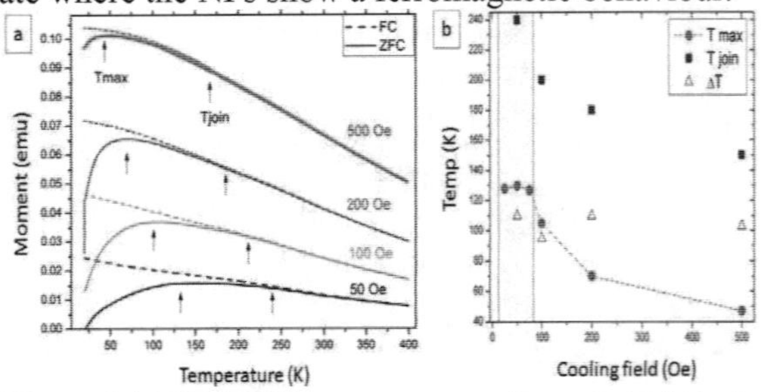

Figure 3 (a) ZFC / FC measurements of sample B (8 nm) at 50, 100, 200 and 500 Oe. (b) Graphs of T_{max} and T_{join} as functions of cooling field value.

Figure 3 shows Zero Field Cooling (ZFC) and Field Cooling (FC) curves for sample B. FC and ZFC curves were measured at different H values from 50 to 500 Oe. All the ZFC curves exhibit a peak, indicating a transition from a magnetically blocked state (at low temperature) to a superparamagnetic state (at high temperature). The peak position T_{max} varies with the magnitude H of the external field, as plotted in Figure 3b. When $H \leq 75$ Oe, T_{max} reaches an optimal value of 130 K, which we ascribe to the blocking temperature T_B. The FC and ZFC curves do not join at T_{max} but at a higher temperature, T_{join}. The gap $\Delta T = T_{join} - T_{max}$ is on the order of 100K for sample B, indicating strong magnetic couplings between the NPs. Figure 4(a-c) shows ZFC / FC curves measured at 100 Oe, for the three NP samples. As expected, T_{max} increases significantly when the NP size increases. Sample A exhibits the lowest $T_{max} = 28$ K, and sample C the highest $T_{max} = 170$ K, as reported in Table 1. Figure 4(d-e) shows magnetization loops measured at 400 K and 20 K for the three NP sizes. The larger 11 nm NPs align more easily than the smaller 5 nm NPs. At 20K, the 5 nm NPs exhibit very little hysteresis, indicating a pure superparamagnetic behaviour, while the larger NPs exhibit a significant hysteresis, with an increased coercivity $H_c \sim 235$ Oe (from ~0 at 400K), indicating ferromagnetic-like couplings.

The magnetic behavior of our 5 to11 nm Fe_3O_4 NPs agrees with data measured on other Fe_3O_4 NPs by other groups.[6-8] Some models predict that, due to surface effects, H_c should increase when the NP size decreases.[9,10] We found that H_c actually decreases when the NP size decreases from 11 to 5 nm (see Table 1). Such trend is also observed by Guardia et al. [8] on 6 to 20 nm NPs, and by Goya et al. [6], on 5 to 150 nm Fe_3O_4 NPs. The consistency of the results, obtained on different NP samples, suggests that surface spin-disorder and resulting magnetic anisotropy are strongly reduced by the presence of oleic acid ligands covalently bonded to the surface.[11] With a ligand shell, the NPs exhibit bulk-like magnetic anisotropy. Our ZFC/FC results, showing that T_{max} drastically increase with NP size, also agree with other reported findings on 5 to 20nm NPs.[11,12] The discrepancies between the observed T_{max} from the various reports may be due to different particle chemical environments. Also, our observed increase of magnetic couplings (where $T_{join} > T_{max}$) for 8 and 11nm NPs compared to the 5nm NPs agrees with other measurements and is attributed to increased magnetic moment.[6]

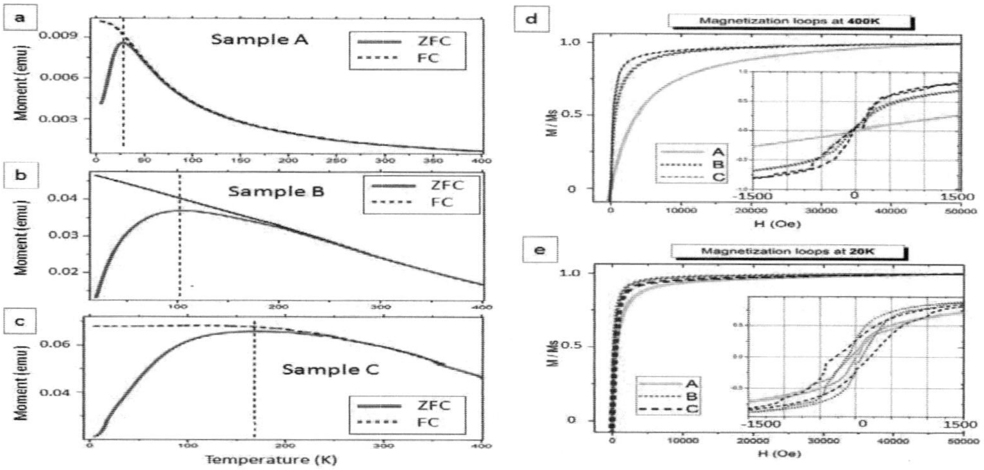

Figure 4. (a-c) ZFC / FC measurements of sample A (5 nm), B (8 nm) and C (11 nm) measured at 100 Oe. (d) Magnetization curves for the three samples, measured at 400K and (e) at 20 K. Insets: close-up on hysteresis.

Table 1. Structural and magnetic parameters for samples A, B and C.

	Particle size		ZFC/FC curves		Magnetization loops			
	TEM (nm)	XRD (nm)	T_{max} (K)	$\Delta T^{\dagger}$ (K)	H_c at 400K (Oe)	H_c at 20K (Oe)	$H_{0.95}$ (Oe) at 400K*	$H_{0.95}$ (Oe) at 20K*
Sample A	5.3 ± 0.7	5.8 ±1.7	28	~ 10	~ 0	~ 25	29,575	14,200
Sample B	8.1 ± 1.7	8.5 ±2.9	130	~ 100	~ 20	105	13,100	3900
Sample C	11.3 ± 2.5	11.0 ± 4.6	170	~ 50	50	235	9145	10,272

$^{\dagger}\Delta T$ is the difference between T_{max} and the joining point for the ZFC and FC curves (above T_{max})
** H_s is measured at the shoulder of the magnetization curve, preceding saturation*

In conclusion, we have measured the magnetic behavior of 5 to 11 nm Fe_3O_4 NPs that were prepared by thermal decomposition of an iron precursor and coated with oleic acid. We found that the magnetic properties of the NPs are greatly impacted by the particle size. Magnetization curves indicate a superparamagnetic behavior at high T with very little hysteresis. Moderate hysteresis that increases with the NP size appears at low T once the NPs are in a blocked state. ZFC/ FC measurements confirm the superparamagnetic behavior and show a transition to a magnetically blocked state at low temperature for all the NP sizes. The peak temperature T_{max} drastically increases with the NP size.

We would like to thank Jeffrey Farrer for help with TEM imaging, and Branton Campbell for helpful discussions about the Fe_3O_4 crystallographic structures. This work has been supported by ORCA and MEG grants from Brigham Young University.

References

[1] Frey N.A, Peng S, Cheng K and Sun S, 2009 *Chem. Soc. Rev.* **38**, 2532
[2] Duguet E, Vasseur S, Monet S and Devoisselle J.M, 2006 *Nanomedicine* **1** (2) 157
[3] Park J et al., 2004 *Nature Materials*, **3**, 891-895
[4] Jana N.R, Chen Y, and Peng X, 2004 *Chemistry of Materials* **16**, 3931-3935
[5] Altavilla C, Ciliberto E, Gatteschi D, and Sangregorio C, 2005 *Advanced Materials* **17**, 8
[6] Goya G.F, Berquo T.S, Fonseca F.C and Morales M.P, 2003 *J. Appl. Phys.* **94** (5) 1
[7] Tartaj P et al., 2001 *Adv. Mater.* **13** 1620.
[8] Guardia P et al., 2007 *J. Magn. Magn. Mater.* **316** e756–e759
[9] Tartaj P, et al.,2003 *J. Phys D* **36** R182.
[10] Blanco-Mantecon M, et al., 2006 *J. Magn. Magn. Mater.* **296** 124
[11] Batlle X et al., 2011 *J Appl. Phys.* **109**, 07B524
[12] Figueroa A.I, et al., 2013 *Nanotechnology* **24**, 155705

Journal of Physics: Conference Series **521** (2014) 012005 doi:10.1088/1742-6596/521/1/012005

Microwave-assisted switching: optimal microwave field for a nanomagnet with surface anisotropy

N. Barros and H. Kachkachi

PROMES-CNRS and Universit de Perpignan Via Domitia, 52 Avenue Paul Alduy, 66860 Perpignan Cedex, France.

Abstract. We compute the optimal microwave field that induces the magnetization reversal of a nanomagnet modeled as a macrospin, while minimizing the total injected energy. Two complementary approaches, numerical and analytical, are presented. We then optimize the microwave field for a nanomagnet with a mixed uniaxial and cubic anisotropy. The optimal microwave fields are found to be modulated both in amplitude and frequency. We find that the cubic anisotropy, which is mainly due here to the surface contribution, has a major effect on the frequency profile, related to the variation of the effective field.

1. Introduction

Magnetic recording is a key technology in the field of high density information storage. In order to increase thermal stability small nanoparticles with high anisotropy may be used. However, high fields are then needed to reverse the magnetization but these are difficult to achieve in current devices. To overcome this so called magnetic recording trilemma, switching can be assisted by a microwave (MW) field. In 2003 Thirion et al.[1] showed that the combination of a DC applied field, well below the Stoner-Wohlfarth switching field, with a small MW field pulse can reverse the magnetization of a nanoparticle. Further experimental and theoretical studies proved that this process is more efficient with a chirped MW field [2]. Indeed, the MW field frequency can then synchronize with the magnetization precession, so that the energy pumping between the MW field and the magnetization can occur through a resonant process.

The aim of this work is to determine the optimal MW field that allows for the magnetization reversal of a nanomagnet while minimizing the total injected energy. We consider a single nanomagnet with homogeneous or nearly collinear magnetization that can be modeled as a macrospin with a mixed uniaxial and cubic anisotropy. The cubic contribution may stem from the cubic magneto-crystalline anisotropy or from surface effects as was shown in Ref. [3]. The study is carried out at zero temperature.

After introducing the model used to describe the dynamics of the magnetization, we present the two complementary approaches that we developed to optimize the MW field. Finally, we numerically compute the optimal MW field for a nanomagnet with a mixed anisotropy and investigate the effect of the cubic contribution.

2. Model and dynamics equation

The magnetic sate of nanoparticle of volume V is modeled by a macrocopic magnetic moment (macrospin) with constant magnitude $\mathbf{M} = M_S\mathbf{m}$, where M_S is the saturation magnetization and $\|\mathbf{m}\| = 1$. The nanomagnet is characterized by a mixed-anisotropy energy comprising a uniaxial term along the z axis and a cubic term. In the presence of a DC magnetic field $\mathbf{H}_{\mathrm{DC}}$, the energy density of the nanomagnet can then be written as

$$E_0/V = -K_2 m_z^2 + \frac{K_4}{2}\left(m_x^4 + m_y^4 + m_z^4\right) - \mathbf{m}\cdot\mathbf{H}_{\mathrm{DC}}. \tag{1}$$

Introducing the anisotropy field $H_{\mathrm{an}} = 2K_2/\mu_0 M_S$, the reduced field $\mathbf{h}_{\mathrm{DC}} \equiv \mathbf{H}_{\mathrm{DC}}/H_{\mathrm{an}}$ and the cubic anisotropy parameter $\zeta \equiv K_4/K_2$, we define the reduced energy density of the nanomagnet by

$$\mathcal{E}_0 \equiv \frac{E_0}{2K_2V} = -\frac{m_z^2}{2} + \frac{\zeta}{4}\left(m_x^4 + m_y^4 + m_z^4\right) - \mathbf{m}\cdot\mathbf{h}_{\mathrm{DC}}. \tag{2}$$

The trajectory of the magnetization is described by the Landau-Lifshitz-Gilbert (LLG) equation

$$\left(1+\alpha^2\right)\frac{d\mathbf{m}}{d\tau} = -\mathbf{m}\times\mathbf{h}_{\mathrm{eff}} - \alpha\mathbf{m}\times\left(\mathbf{m}\times\mathbf{h}_{\mathrm{eff}}\right) \tag{3}$$

where α is the damping parameter of the nanomagnet, $\tau \equiv (\gamma H_{\mathrm{an}})\,t$ is the reduced (dimensionless) characteristic time, $\gamma \simeq 1.76\times 10^{11}\mathrm{rad/(T.s)}$ the gyromagnetic ratio, and $\mathbf{h}_{\mathrm{eff}} = -\partial\mathcal{E}_0/\partial\mathbf{m}$ is the static effective field.

In the presence of a reduced microwave field $\mathbf{h}_{\mathrm{MW}}(\tau)$, in Eq. (3) $\mathbf{h}_{\mathrm{eff}}$ is replaced by $\mathbf{h}_{\mathrm{eff}} + \mathbf{h}_{\mathrm{MW}}(\tau)$. Then, the energy variation of the system is given by

$$\frac{\partial\mathcal{E}}{\partial\tau} = -\frac{\alpha}{1+\alpha^2}\left\|\mathbf{m}\times\left(\mathbf{h}_{\mathrm{eff}} + \mathbf{h}_{\mathrm{MW}}(\tau)\right)\right\|^2 - \mathbf{m}\cdot\frac{d\mathbf{h}_{\mathrm{MW}}(\tau)}{d\tau}. \tag{4}$$

Note that, contrary to a DC field which always tends to decrease the energy of the system, a MW field can be a source of energy. If properly designed, it can induce the switching of the magnetization from one minimum to another.

3. Optimization of the microwave field

Setting the magnetization initially in an energy minimum we seek the optimal MW field that triggers its reversal to another minimum in a given time τ_f while minimizing the total injected energy, defined by $E_{\mathrm{inj}} = \int_0^{\tau_f}\left\|\mathbf{h}^2_{\mathrm{MW}}(\tau)\right\| d\tau$. This amounts to finding the best compromize between the MW field pulse intensity and duration, so that the subsequent heating of the system will be minimized. For that, we proposed two complementary approaches [4; 5].

The numerical approach [4] is based on the optimal control theory. We define the cost functional

$$\mathcal{F}\left[\mathbf{m}(\tau), \mathbf{h}_{\mathrm{MW}}(\tau)\right] = \frac{1}{2}\left\|\mathbf{m}(\tau_f) - \mathbf{m}_f\right\| - \frac{\eta}{2}\int_0^{\tau_f}\left\|\mathbf{h}^2_{\mathrm{MW}}(\tau)\right\| d\tau \tag{5}$$

where $\mathbf{m}(\tau_f)$ is the magnetization reached at the final time τ_f, $\mathbf{m}_f$ is the magnetization at the second minimum (the target), and η is a numerical control parameter. Introducing a Lagrange parameter, we then minimize this cost functional along the trajectory given by the driven LLG equation. Because of the non linearity of the problem this is done numerically using the Conjugate Gradient technique augmented by a Metropolis algorithm.

In the analytical approach [5], we first determine the critical MW field that exactly compensates for the effects of damping, thus maintaining the magnetization precession at any time. Assuming that the latter and the optimal MW field we seek have a similar shape, we find that the optimal MW field is given by

$$\mathbf{h}^{\mathrm{opt}}_{\mathrm{MW}}(\tau) = \frac{2\alpha}{1+\alpha^2}\left[-\mathbf{m}\times\mathbf{h}_{\mathrm{eff}} + \alpha\mathbf{m}\times\left(\mathbf{m}\times\mathbf{h}_{\mathrm{eff}}\right)\right]. \tag{6}$$

In the presence of this MW field, the magnetization trajectory can be described by the LLG equation with a negative damping parameter, which corresponds to an "amplified" precession with a frequency $\omega = \|\mathbf{h}_{\text{eff}}\| / (1 + \alpha^2)$.

Both approaches have been favorably compared in the case of a nanomagnet with uniaxial anisotropy ($\zeta = 0$). The optimal MW field is found to be modulated both in amplitude and frequency, and synchronized at any time with the magnetization precession. Its role is to drive the magnetization from the first minimum to a saddle point, then the damping induces the relaxation down to the second minimum. Moreover, we showed that the total injected energy E_{inj} is proportional to the damping parameter and to the static energy barrier between the saddle point and the first minimum.

4. Effect of the cubic anisotropy

Here we drop the DC field and assume that the nanomagnet is characterized by a mixed anisotropy energy. The cubic contribution can result from the magneto-crystalline anisotropy or stem from surface effects. We will restrict our study to moderate values of ζ ($-1 < \zeta < 1$) that have been observed in real nanoparticles [6]. The potential energy surface presents two minima at $\mathbf{m} = (0, 0, 1)$ (up state) and $\mathbf{m} = (0, 0, -1)$ (down state) separated by four saddle points with the same energy located in the xy plane, along the cube diagonals if $\zeta > 0$ or along the x, y axes if $\zeta < 0$. The magnetization is initially set in the up state. For $\zeta \neq 0$ the time evolution of the optimal MW field can not be expressed analytically. Hence, only the numerical approach has been used to optimize the MW field.

The optimal MW field was firstly determined for $\zeta = 0.3$ (Fig. 1). It is found to be modulated both in amplitude and frequency, and to lie mainly in the xy plane. Moreover, it is synchronized with the magnetization. It drives the magnetization away from the up state and to climb up the potential well, until it reaches the saddle point defined by $\mathbf{m} = \left(1/\sqrt{2}, -1/\sqrt{2}, 0\right)$ at $\tau \approx 680$. Next, the MW field vanishes and the magnetization, driven by damping, relaxes towards the down state. More runs of the program with the same parameters rendered nearly superposable MW fields, the magnetization trajectory crossing randomly any of the four saddle points.

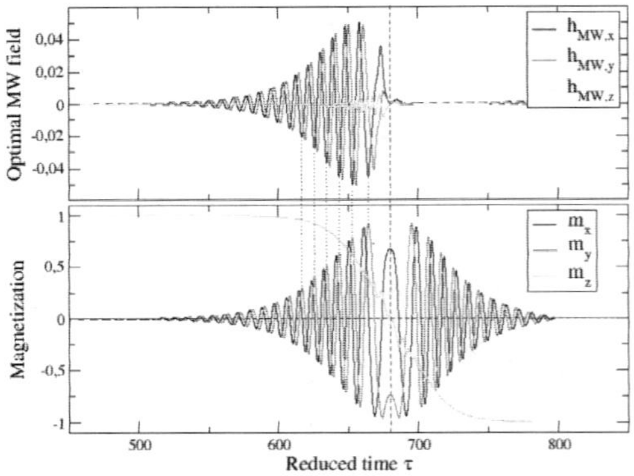

Figure 1. Optimal MW field (upper panel) and corresponding magnetization trajectory (lower panel) obtained numerically for $\zeta = 0.3$. The blue dotted line indicate the crossing of the saddle point. Parameters: $\alpha = 0.05$, $t_f = 800$, $\eta = 0.01$, number of points $N = 15000$.

We optimized the MW fields for different values of ζ (Fig. 2). As before, the optimal fields are found to be chirped and synchronized with the magnetization precession. As can be seen, the main effect of ζ on the optimal MW field is to modify its frequency profile. This can be related to the variation of the effective field with ζ. When the magnetization goes from the up state to one of the saddle points, the effective field magnitude $\|\mathbf{h}_{\text{eff}}\|$ varies from $1 - \zeta$ to $|\zeta|$. For low or negative values of ζ, we thus expect the MW field frequency to decrease continuously during the pulse. On the other hand, for high values of ζ the frequency should be low at the beginning and increase during the pulse. This is what is observed for example for $\zeta = 0.8$.

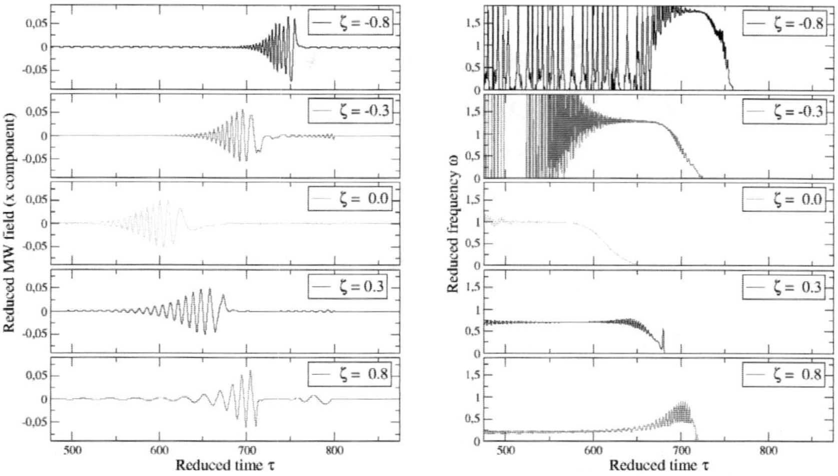

Figure 2. Optimized MW fields obtained for different values of ζ (left) and corresponding frequencies (right). Same parameters as Fig. 1.

5. Conclusion

In this work, we computed the optimal MW that induces the switching of a nanomagnet modeled as a macrospin in an effective potential comprising a mixed anisotropy and a Zeeman term. After presenting the two complementary approaches that we developed, we used the numerical approach to optimize the MW field for different values of the cubic anisotropy parameter ζ. We find that the optimal field is modulated both in amplitude and frequency, and that the frequency profile strongly depends ζ. This result can be used to design MW fields to induce switching in real nanoparticles that often exhibit a complex energy anisotropy due to *inter alia* surface effects. From a fundamental point of view, it is also interesting to consider higher values of ζ, since secondary minima can appear, creating many possible trajectories for the magnetization. Finally, it will be necessary to introduce the effects of temperature and determine if they can favor or prevent the microwave-assisted switching.

References

[1] Thirion Ch, Wernsdorfer W and Mailly D 2003 *Nat. Mat.* **2** 524.

[2] Okamoto S, Kikuchi N and Kitakami O 2008 *Applied Physics Letters* **93** 142501.

[3] Garanin D A and Kachkachi H 2003 *Phys. Rev. Lett.* **90**(6) 065504.

[4] Barros N, Rassam M, Jirari H and Kachkachi H 2011 *Phys. Rev. B* **83**(14) 144418.

[5] Barros N, Rassam H and Kachkachi H 2013 *Phys. Rev. B* **88**(1) 014421.

[6] Jamet M, Wernsdorfer W, Thirion C, Mailly D, Dupuis V, Mélinon P and Pérez A *Phys. Rev. Lett.* **86** 4676; Jamet M *et al.* 2004 *Phys. Rev. B.* **69** 024401.

Understanding dynamics of interacting magnetic nanoparticles: from the weak interaction regime to the collective superspin glass state

D. Fiorani* and D. Peddis

ISM-CNR, Area della Ricerca Roma 1, Via Salaria km 29.300, Monterotondo Scalo (RM), Italy

Email: dino.fiorani@ism.cnr.it

Abstract.

A brief review of the investigations on interparticle interaction effects (weak and strong interactions regimes) is reported with a special emphasis on the critical behaviour in the static and dynamical properties and on the out of equilibrium dynamics of the collective superspin glass (SSG) state. The results of the investigation of aging and memory effects in two peculiar SSG systems are discussed: $MnFe_2O_4$ nanoparticles, in powder form, with very strong dipolar interactions, and a diluted system, consisting of a film of Co particles dispersed (5 % volume fraction) in a Mn matrix, where interparticle interactions are mainly mediated by the matrix.

1. Introduction

Since the pioneering work of Louis Néel in the 1940s[1], nanoparticle magnetism has been representing one of the most attractive and challenging topics in both fundamental[2,3] and applied magnetism (magnetic recording, biomedicine, energy, sensors….)[4–7]. The topic, having an interdisciplinary character, has been evolving continuously and in particular the progress was enormous in the last two decades, with the discovery of new phenomena and the opening of new perspectives for applications in different fields. They come from the control of the interplay between intrinsic intraparticle properties (finite size and surface effects) and interparticle interaction effects. The latter is responsible for the existence of different types of collective magnetic states and has a dominant role in technological applications. Moreover, the capability of engineering, by advanced chemical and physical methods, organized arrays of nanoparticles and ordered structures of single domains nanodots[8,9], has been further expanding the applications in nanotechnology. This paper is aimed at briefly reviewing the progress, since the first International Conference on Fine Particle Magnetism (ICFPM – Rome, 1991)[10], in the understanding of interparticle interactions effects leading to the collective superspin glass (SSG) state. The results of the investigation of out of equilibrium dynamics in two peculiar SSG systems will be discussed: $MnFe_2O_4$ nanoparticles, in powder form, with very strong dipolar interactions, and a diluted system consisting of a film of Co particles embedded (5 % volume fraction) in a Mn matrix, where the interactions are mainly mediated by the matrix.

2. Interparticle interactions effects

2.1 Weak interparticle interactions regime

Until the early 90s, the study of the effect of interparticle interactions was restricted to the weak interactions regime, defined as the regime of an assembly of magnetic nanoparticles where the interparticle dipolar interaction energy (E_{int}), contributing to the total effective anisotropy energy (E_{tot}), is much lower than the individual particle anisotropy energy (E_a). The superparmagnetic (*SP*) model basically still holds, provided that the actual contribution of interparticle interactions to the total anisotropy energy is taken into account ($E_{tot} = E_a + E_{int}$). In such conditions, the effect of interparticle interactions was considered just a perturbation to the *SP* state, leading to the so called "*modified SP*" regime, where the static and dynamical properties of individual particles are partially modified by the interaction with the neighbouring particles.

In the pure *SP* regime, where interparticle interactions are absent, the static properties of the particles are described by the Langevin function, like for paramagnets. Allia et al. [11] proposed a modified Langevin function to account for the presence of weak interparticle interactions:

$$M = N\mu_n L\left(\frac{\mu_n H}{k_B(T - T^*)}\right) \qquad (1)$$

Where μ_n is the nanoparticle moment, N is the number of moments per unit volume, L is the Langevin function and T^* is fictive temperature (a positive definite quantity), which basically measures the magnitude of the dipolar interaction and depends on structural parameters, such as magnetic nanoparticles density, and on saturation magnetization. For a given nanoparticles system a T* value is derived at each temperature (being T* weakly dependent on temperature because it contains the sample's magnetization).

The model satisfactorily explains the main features of the M vs H curves. Such regime was defined by the authors as an "*interacting superparamagnet*", i.e. a SP regime where the dipolar interactions are no longer negligible.

The dynamical properties for an assembly of non interacting nanoparticles with uniaxial symmetry were described by the Néel model[1], which predicts a temperature and anisotropy energy dependence of the relaxation time (τ) according to an Arrhenius law. This law implies a $Tln t/\tau_o$ scaling in the whole *SP* regime[12]. The effect of interparticle interactions on the dynamical properties was described modifying the Arrhenius law for SP systems and replacing it by the phenomenological Fulcher law:

$$\tau = \tau_0 \exp\left(\frac{K_a V}{k_B(T - T_0)}\right) \qquad (2)$$

where the effect of weak interparticle interactions is accounted by the temperature T_0, giving an estimation of their strength[13].

A more general description of the dynamical properties was given by Dormann et al[14] with a model describing the effect of magnetic dipolar interactions on the relaxation time of the particle moments. A disordered assembly of non-identical particles with volume distribution was considered. This implies a distribution of relaxation times and a distribution of energy barriers, which depends on the particle volume and on the dipolar interactions. Since the particle moments fluctuate with different τ, even the angle between two non identical particles fluctuates with time. Thus, it is not possibile to calculate the effect of the dipolar interactions strength in static conditions. To model the effect of interpaticle interactions in dynamical conditions, the Boltzman statistics was used introducing a modified Langevin function containing the interparticle interactions dipolar energy, which adds to the individual particle anisotropy energy. Considering the approximation $E_{int} << E_a$, the interaction energy can be written as:

$$E_{\text{int}} = \left(\frac{\mu}{V}\right)\Sigma_j b_j L\left(\frac{\mu^2 a_j}{VK_B T}\right) \quad (3) \qquad a_j = V\frac{2\cos 2\alpha_j}{d^3} \quad (3b)$$

Where a_j, b_j, α_j are parameters that define the spatial arrangement of nanoparticles. The frequency dependence of the blocking temperature (T_b) is well reproduced by the model in a very large time window. Interparticle interactions modify the energy barrier, leading to an increase of T_b, in agreement with experimental observations[15,16].

2.2 Strong interparticle interactions regime

In the "*strong interactions*" regime, the interparticle interaction energy is much larger than the individual particle anisotropy energy ($E_{int} >> E_a$). Moving from the weak to strong interparticle interactions regime, a crossover occurs from the blocking of particle moments to a collective magnetic state where particle moments lose their individuality. This is observed in concentrated frozen ferrofluids [17,18] and in concentrated granular systems where single domain particles are dispersed in a non magnetic matrix[19]. Two sub-regimes are distinguished in the strong interactions regime, with increasing the particle concentration: the superspin glass (SSG) regime and the superferromagnetism (SFM) regime. The difference in the two sub-regimes is in the strength of interparticle interactions, which is moderate in SSG and stronger in SFM, and in the symmetry of interactions, leading to a disorder state or to an ordered one, in SSG and SFM, respectively. For large particle concentration, but prior to physical percolation, the interparticle interactions become much stronger than in SSG and finally they can lead to the SFM state, characterized by FM domains of supermoments of particles, i.e. domains where ferromagnetic correlations are between supermoments of individual particles, instead that between atomic moments[20].

2.2.1 Superspinglass

The investigation of similarities between a SG system and strongly interacting nanoparticle systems has been receiving a growing attention since the early 90s [16,17,21–29]. In a particle system with sufficiently narrow size distribution, sufficiently strong interparticle interactions, randomness in the distribution of particle positions and of anisotropy axes orientations, a SSG state can be observed, characterized by a collective random freezing of particle moments below a characteristic glass temperature (T_g). Both the required ingredients of the SG state are also present in the SSG one, i.e. randomness and frustration of interactions between particle moments. A SSG behaviour was recently observed in a random-close-packed (volume fraction 67%) ensemble of highly monodisperse nanoparticles (γ-Fe_2O_3)[30]. Being the collective freezing between particle moments rather then between atomic spins, the dynamics in SSG is much slower, as the microscopic flip time of one superspin (in the order of 10^{-9}s at room temperature; up to $10^{-6;}$ in the frozen state at low temperature) is much longer than an atomic spin flip time (in the order of 10^{-12} s). The growth of dynamical correlation length is slower in SSG and thus the slower dynamics of SSG is of particular interest because a much shorter time scale becomes experimental accessible. Nevertheless, theoretical models developed for atomic SG have so far succeeded in describing many aspects of the SSG static and dynamical properties. It was found, by frequency dependent ac susceptibility and ZFC relaxation measurements, that the growth of supermoments correlation length in a disordered SSG state of a frozen ferrofluid sample follows the same physical law as for Heisenberg SG [17].

SSG exhibits the same type of features of SG both in the static and dynamic properties. The temperature dependence of the low field susceptibility is characterized by a maximum of the zero-field-cooled (ZFC) susceptibility and small maximum of the field cooled (FC) susceptibility, followed by a plateau with decreasing temperature. Ageing and memory effects, typical of SG behaviour, are also observed in SSG. The basic analogy is the critical behaviour in both static and dynamical properties, which represent the real proof of the existence of the special type of transition at T_g to the SG and SSG state.

A critical behaviour in the temperature dependence of the non linear susceptibility χ_2, diverging at T_g, was observed in SSG systems (e.g. Fe_2N[27], γ-Fe_2O_3[31] Fe-C[32] nanoparticles) A critical behaviour was also observed in the temperature dependence of the relaxation time with decreasing temperature, diverging at T_g, according to a power law, with critical exponents having values quite similar to those reported for SG (zν = 8-10) [31–33] (z is the dynamical critical exponent; ν is the critical exponent driving the correlation length, ξ^{ν}).

$$\tau = \tau_0 \left(\frac{T}{Tg - 1} \right)^{-z\nu} \quad (4)$$

The zν value was found to be dependent on the anisotropy of the SG system, as shown for Ising ($Fe_{0.5}Mn_{0.5}TiO_3$; zν = 10.5) and Heisenberg like ($CdCr_{1.7}In_{0.3}$; zν = 7) SG[34].Another evidence of SG like properties is the observed dynamic scaling, i.e. the collapse onto a master curve of the values of the imaginary component of the dynamic susceptibility (χ'') measured at different frequencies and temperatures[31,32].

The dynamical properties below T_g are characterized by ageing and memory effects, which are manifestation of an out of equilibrium state, associated to the onset of a random collective state of particle moments. As in SG, even in SSG it was observed a slowing down of the relaxation of the ZFC magnetization with increasing the time spent (waiting time, t_w) at a constant temperature below T_g, before the application of the magnetic field [21]. This reflects the slow evolution of the system towards an equilibrium configuration, during the aging, starting at the time of the quench below T_g. The aging effect was satisfactorily explained both by the model of hierarchical system of energy minima in the phase space, by Parisi[35] and the droplet model of Fisher and Huse[36]. Ageing is a manifestation of a slow growth of a dynamical correlation length, when the system is aged in the frozen state, developing in the same way as the correlations between the spins of an atomic spin glass. The ageing phenomenon is absent in the most diluted samples revealing that this phenomenon is a manifestation of a collective dynamical behaviour.

Memory effects were also observed in the ZFC magnetization curve, as in SG, following the so called "*stop and wait*" experimental protocol during the ZFC process: the ZFC cooling down is interrupted at a given temperature below T_g, the system is maintained at this temperature for a certain time and then the cooling continues down to the lowest temperature. When the magnetization is measured on warming up, lower values of the magnetization with respect the values of the reference curve, measured without interruption in the ZFC process, are observed in the temperature region where the previous ZFC cooling process was stopped. This phenomenon is another manifestation of ageing, which occurs during the stop time at the stop temperature before cooling down to the lowest T, and it reveals that the system remembers the stopping procedure[22,36].

3.Examples of SSG systems

3.1 $MnFe_2O_4$ nanoparticles

$MnFe_2O_4$ nanopowders have been synthesized by coprecipitation of Fe^{3+} and Mn^{2+} from water-in-toluene reverse micelle system and subsequent thermal treatment at 320°C. The mean crystallite size obtained by XRD analysis is ~2 nm, whereas the value obtained by powder specific area (278 m^2/g) is about 4 nm. This suggests that the particles consist of a few aggregated crystallites[38,39].

ZFC and FC magnetization measurements, recorded with an applied field of 100 Oe (figure 1a) show the typical SSG behaviour, with a sharp maximum of both curves at Tg = 45 K. A critical divergence of the relaxation time, according to a power law (zν = 8.6), was found by the analysis of the frequency dependence of T_g by using ac susceptibility measurements. The non equilibrium dynamics below T_g has been investigated by memory and rejuvenation experiments. A ZFC reference curve was recorded according to the usual procedure. The memory curves (Fig. 1b) were measured in the same way, but after maintaining the sample in zero field for 3 h at 20 K (T/T_g = 0.45) and 33 K (T/T_g = 0.45). A reduced magnetization was observed at the stop and wait temperatures, the effect being larger at the higher temperature.

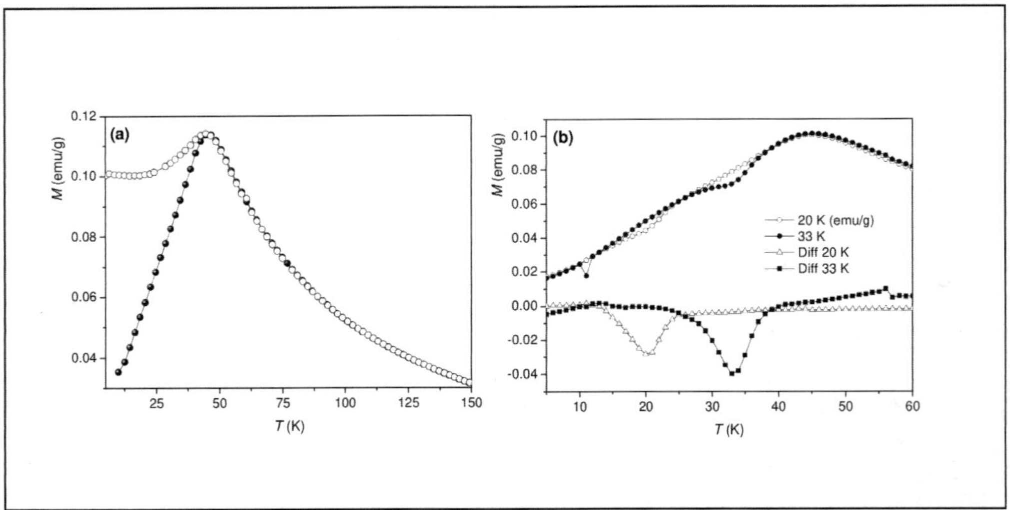

Figure 1: (a) FC (empty circles) and ZFC (full circles) magnetization of $MnFe_2O_4$ nanoparticles in an applied field of H=100 Oe; (b) upper curves: ZFC-memory curves for a stop-and-wait of 10000 sec at 0.45 T/T_f (T= 20 K; empty circles) and at 0.75 T/T_f (T= 33 K; full circles); lower curves: difference, multiplied by a factor 10, between ZFC-reference and ZFC-memory curve (full circles in panel a) at 0.45 T/T_f and at 0.75 T/T_f.

3.2 Co nanoparticles embedded in antiferromagnetic Mn matrix

Co particles embedded in Mn matrix (Co@Mn) with ~5% volume fraction (VF) were prepared in thin film form by co-deposition using a gas aggregation cluster source and a molecular beam epitaxy (MBE) source[39]. The cluster source produces a log-normal distribution of particle sizes, with a mean size of 1.8 nm, as measured in situ by an axially mounted quadrupole filter. In order to clarify the role of the antiferromagnetic matrix, a reference sample of Co nanoparticle embedded in diamagnetic Ag matrix with the same volume fraction (Co@Ag) was prepared[40,41]. Figure 2a reports ZFC and FC susceptibility for Cobalt nanoparticles dispersed in Mn and Ag matrix. The comparison provides a clear evidence of an independent blocking of particle moments in Co@Ag and of a correlated freezing of particle moments, SSG type, in (Co@Mn). Indeed, in Co@Ag the continuous increase of χ_{FC} with decreasing temperature, indicates that interparticle interactions are very weak. On the other hand, in Co@Mn the FC susceptibility shows, with decreasing temperature, a sharp maximum at 70 K followed by a plateau below 30 K. This landscape is confirmed by AC susceptibility measurements. In Co@Ag, the frequency dependence of T_{max} is well described by an Arrhenius law, whereas in Co@Mn by a power law ($z\nu \cong 8.2$)[41]. The Co@Mn sample exhibits a non equilibrium glassy dynamics, as shown the by the stop and wait experiment using the ZFC protocol (inset figure 2a). In Fig. 2b, Monte Carlo (MC) simulation of ZFC curves for different waiting times t_w = 300, 1000, 3000, and 10000 s are reported on a normalized temperature scale. The differences between the reference and memory curves for different t_w are reported in the inset of Fig. 2b. For a logarithmic spacing of waiting times as those used in the experiment, equally spaced depths in the measured memory dips were observed. In this dilute system, the interface exchange coupling between Co ferromagnetic particles and the antiferromagnetic Mn matrix plays a very important role in transmitting long range interparticle interactions. Actually, due to a partial alloying (the existence of Co_xMn_x alloy was shown by EXAFS data) the interface coupling is between the ferromagnetic Co core and the surrounding antiferromagnetic Co_xMn_x shell, consistently with the very low Co moment measured by XMCD spectra [40]. MC Simulations were also performed for a 10% VF sample, larger memory effect were experimentally observed. This indicates that both dipolar interparticle interactions and the interface exchange coupling with the Mn matrix contribute to stabilize a SSG phase and its dynamical behaviour. Simulations including both contributions reproduce well the memory effect for different t_w. Simulations without including dipolar interactions still show the memory effect, although weaker, confirming the role played by the matrix. The interface exchange coupling between the FM and AFM phase is also responsible for the observed Exchange Bias (EB) effect below the temperature at which the FC susceptibility starts to become temperature independent, signalling a complete freezing of particle moments[42-44]. Due to the intrinsic energy structure of SSG, characterized, according to the Parisi's model[35,44], by a hierarchical organization of many energy minima separated by energy barriers of different height, EB involving a SSG phase provides an even better possibility of tuning the coercivity, since it is dependent on the energy barrier distribution [41][45].

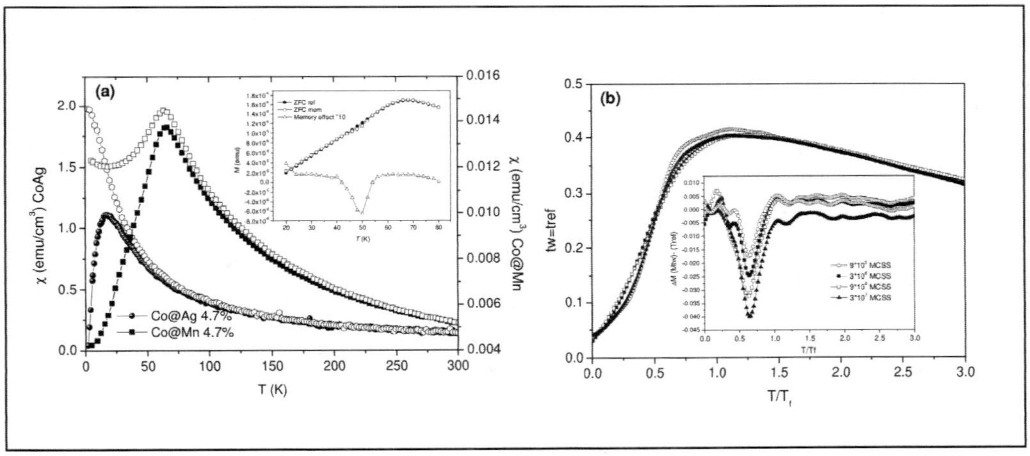

Figure 2: (a) ZFC (full circles/squares) and FC (empty circles/squares) susceptibility for Co@Ag (circles) and Co@Mn (squares) samples with Co VF 5%; Inset, upper curve: ZFC-reference curve (full circles) and ZFC-memory curve (empty circles squares: for a stop-and-wait of 10000 s at $T/T_f \approx 0.8$ (T= 50 K); lower curve: difference, multiplied by a factor 10, between ZFC-reference and ZFC-memory curve (empty triangles). (b) Monte Carlo simulation of ZFC-FC reference curve ($t_{ref} = 3\times10^3$ MCSS) and ZFC-memory curves after stop-and-wait for different waiting times at $T/T_f = 0.64$ for 5 % volume fraction and H = 0.20 JC/gμB; inset: difference between ZFC-reference and ZFC-memory curve as a function of temperature for different waiting times t_w.

4. Conclusions

Interparticle interaction effects have been reviewed with a special emphasis on the analogies between the static and dynamical properties of a collective frozen disordered state of particle supermoments (SSG state) and those of atomic spins (SG state). Apart from the much slower dynamics in SSG, the same phenomenology as in Heisenberg SG is observed: temperature dependence of the ZFC/FC susceptibility curves, critical divergence at T_g of the non linear susceptibility and of the relaxation time with comparable critical exponents; dynamic scaling, ageing and memory effects. The results of the investigation of out of equilibrium dynamics in two SSG systems have been discussed: $MnFe_2O_4$ nanoparticles, in powder form, with very strong dipolar interparticle interactions, and a diluted system, consisting of a film of Co particles dispersed (5% volume fraction) in a Mn matrix, where the antiferromagnetic matrix, interface coupled with the ferromagnetic particles, plays an important role in transmitting long range interparticle interactions.

References

[1] L. Néel, Ann. Gèophysique 5 (1949) 99.

[2] J.L. Dormann, D. Fiorani, E. Tronc, in:, wiley (Ed.), Adances Chem. Phys., New York, 1997.

[3] S. Morup, M.F. Hansen, C. Frandsen, Magnetic Nanoparticle, 2011.

[4] D. Weller, a. Moser, IEEE Trans. Magn. 35 (1999) 4423.

[5] C.B.M. S. Sun D. Weller, L. Folks and A. Moser, Science (80-.). 287 (2000) 1989.

[6] Q.A. Pankhurst, N. Thanh, S. Jones, J. Dobson, J. Phys. D. Appl. Phys. 42 (2009) 224001.

[7] L. Suber, D. Peddis, in:, Nanomater. Life Sci., Wiley, 2010, p. 431475.

[8] T.R. Albrecht, D. Bedau, E. Dobisz, H. Gao, M. Grobis, O. Hellwig, D. Kercher, J. Lille, E. Marinero, K. Patel, R. Ruiz, M.E. Schabes, L. Wan, D. Weller, IEEE Trans. Magn. 49 (2013) 773.

[9] S. Sun, S. Anders, H.F. Hamann, J.-U. Thiele, J.E.E. Baglin, T. Thomson, E.E. Fullerton, C.B. Murray, B.D. Terris, J. Am. Chem. Soc. 124 (2002) 2884.

[10] J.L. Dormann, D. Fiorani, in:, North-Holland (Ed.), Amsterdam, 1992.

[11] P. Allia, M. Coisson, P. Tiberto, F. Vinai, M. Knobel, M.A. Novak, W.C. Nunes, Phys. Rev. B 64 (2001) 144420.

[12] A. Labarta, O. Iglesias, L. Balcells, F. Badia, Phys. Rev. B 48 (1993) 10240.

[13] S. Shtrikman, E.P. Wohlfarth, Phys. Lett. 85 (1981) 467.

[14] J.L. Dormanni, L. Bessaist, D. Fiorani, J. Phys. C Solid State Phys. 21 (1988) 2015.

[15] D. Fiorani, J.L. Dormann, R. Cherkaoui, E. Tronc, F. Lucari, F. D'Orazio, L. Spinu, M. Nogues, A. Garcia, A.M. Testa, J. Magn. Magn. Mater. 196-197 (1999) 143.

[16] J.L. Dormann, R. Cherkaoui, L. Spinu, M. Noguès, F. Lucari, F. D'Orazio, D. Fiorani, A. Garcia, E. Tronc, J.P. Jolivet, J. Magn. Magn. Mater. 187 (1998) L139.

[17] S. Nakamae, C. Crauste-Thibierge, D. L'Hôte, E. Vincent, E. Dubois, V. Dupuis, R. Perzynski, Appl. Phys. Lett. 101 (2012) 242409.

[18] S. Nakamae, C. Crauste-Thibierge, K. Komatsu, D. L'Hôte, E. Vincent, E. Dubois, V. Dupuis, R. Perzynski, J. Phys. D. Appl. Phys. 43 (2010) 474001.

[19] D. Peddis, C. Cannas, A. Musinu, Piccaluga.G., J. Phys. Chem. C 112 (2008) 5141.

[20] S. Bedanta, W. Kleemann, W. Kleeman, J. Phys. D. Appl. Phys. 42 (2009) 013001.

[21] T. Jonsson, J. Mattsson, C. Djurberg, F.A. Khan, P. Nordblad, P. Svedlindh, Phys. Rev. Lett. 75 (1995) 4138.

[22] S. Sahoo, O. Petracic, W. Kleemann, P. Nordblad, S. Cardoso, P.P. Freitas, J. Magn. Magn. Mater. 272-276 (2004) 1316.

[23] T. Jonsson, P. Nordblad, P. Svedlindh, Phys. Rev. B 57 (1998) 497.

[24] S. Nakamae, C. Crauste-Thibierge, K. Komatsu, D. L'Hôte, E. Vincent, E. Dubois, V. Dupuis, R. Perzynski, J. Phys. D. Appl. Phys. 43 (2010) 474001.

[25] D. Parker, V. Dupuis, F. Ladieu, J.P. Bouchaud, E. Dubois, R. Perzynski, E. Vincent, Phys. Rev. B 77 (2008) 104428.

[26] J.L. Dormann, D. Fiorani, E. Tronc, J. Magn. Magn. Mater. 202 (1999) 251.

[27] H. Mamiya, I. Nakatani, T. Furubayashi, Phys. Rev. Lett. 80 (1998) 177.

[28] M. Sasaki, H. Takayama, H. Mamiya, P.E. Jönsson, Phys. Rev. B 71 (2005) 104405.

[29] H. Mamiya, S. Nimori, M. Ohnuma, I. Nakatani, M. Demura, T. Furubayashi, J. Magn. Magn. Mater. 316 (2007) e535.

[30] J. a. De Toro, S.S. Lee, D. Salazar, J.L. Cheong, P.S. Normile, P. Muñiz, J.M. Riveiro, M. Hillenkamp, F. Tournus, a. Tamion, P. Nordblad, Appl. Phys. Lett. 102 (2013) 183104.

[31] P. Jönsson, T. Jonsson, J.L. Garcmh, P. Svedlindh, J. Magn. Magn. Mater. 222 (2000) 219.

[32] M.F. Hansen, P.E. Jonsson, P. Nordblad, P. Svedlindh, J. Phys. Condens. Matter 4901 (2002).

[33] D. Peddis, D. Rinaldi, G. Ennas, A. Scano, E. Agostinelli, D. Fiorani, Phys. Chem. Chem. Phys. 14 (2012) 3162.

[34] V. Dupuis, E. Vincent, J.-P. Bouchaud, J. Hammann, a. Ito, H. Katori, Phys. Rev. B 64 (2001) 174204.

[35] G. Parisi, Phys. Rev. Lett. 50 (1983) 1946.

[36] D.S. Fisher, D.A. Huse, Phyiscal Rev. B 38 (1988).

[37] S. Sahoo, O. Petracic, W. Kleemann, P. Nordblad, S. Cardoso, P. Freitas, Phys. Rev. B 67 (2003) 214422.

[38] M. Bellusci, B. Aurelio La, S. Luca, P. Franco, P. Antonella, V. Francesca, Polym. Int. 58 (2009) 1142.

[39] M. Bellusci, S. Canepari, G. Ennas, A. La Barbera, F. Padella, A. Santini, A. Scano, L. Seralessandri, F. Varsano, Mariangela Bellusci, A. La Barbera, J. Am. Ceram. Soc. 3983 (2007) 070922001308006.

[40] N. Domingo, A.M. Testa, D. Fiorani, C. Binns, S. Baker, J. Tejada, J. Magn. Magn. Mater. 316 (2007) 155.

[41] C. Binns, N. Domingo, a M. Testa, D. Fiorani, K.N. Trohidou, M. Vasilakaki, J.A. Balckman, A.M. Asaduzzaman, S. Baker, M. Roy, D. Peddis, J. a Blackman, J. Phys. Condens. Matter 22 (2010) 436005.

[42] N. Domingo, D. Fiorani, A.M. Testa, C. Binns, S. Baker, J. Tejada, J. Phys. D. Appl. Phys. 41 (2008) 134009.

[43] M. Vasilakaki, K.N. Trohidou, D. Peddis, D. Fiorani, R. Mathieu, M. Hudl, P. Nordblad, C. Binns, S. Baker, Phys. Rev. B 140402 (2013) 1.

[44] G. Parisi, Phys. Rev. Lett. 43 (1979) 1754.

[45] D. Fiorani, L. Del Bianco, a. Testa, K. Trohidou, Phys. Rev. B 73 (2006) 092403.

Crystalline Order Effects on the Magnetic Properties of Superlattices made of Cobalt Nanocrystals

Isabelle Lisiecki[1] and Sawako Nakamae[2]

[1] Laboratoire des Matériaux Mésoscopiques et Nanométriques, UMR CNRS 7070, Université P. et M. Curie Bât F, 4 Place Jussieu, 75005 Paris, France.

[2] Service de Physique de l'Etat Condensé, CNRS-URA 2464, IRAMIS/DSM/CEA-Saclay, 91191 Gif-sur-Yvette, France

isabelle.lisiecki@upmc.fr, sawako.nakamae@cea.fr

Abstract. We present the magnetic property evolution in long-range ordered fcc supracrystals of Co nanocrystals (NCs) induced mainly by nanocrystallinity. By increasing the nanocrystalline order of individual NCs, ac susceptibility measurements indicate a significant increase in the 'interaction-to-anisotropy' energy ratio, a trend favourable for the formation of SFM state in these 3D supracrystals.

1. Introduction

Materials based on magnetic nanocrystals (NCs) have been exploited in various applications such as ultra-high density magnetic information storage, magnetic sensors, and biomedical field [**1**] for the last few decades. Interestingly, there remain many questions related to the fundamental 'magnetism' of nanoparticles, and we continue to discover their new properties.

Interacting, single-domain magnetic NCs are known to undergo a superparamagnetic (SPM)-to-superspin glass (SSG) transition at sufficiently high concentrations [2]. By further increasing the inter-particle interaction, creating of a long-range ordered state; *i.e.*, superferromagnetic (SFM) state is said to be possible. Together, these magnetic states created by magnetic nanoparticles now form a new branch of magnetism called "Supermagnetism" [3]. Previously, we reported SSG-like magnetic properties of 3D supracrystals characterized by long-range face-centered cubic (fcc) ordering, made of low anisotropy 8 nm-Co NCs [4]. In this paper, we investigate the effect of nanocrystalline order of Co NCs (from almost amorphous to single-domain hcp) on the onset of SSG state in long-range ordered fcc supracrystals.

2. Experimental Section

2.1. Synthesis of cobalt nanocrystals and elaboration of supracrystals

All materials were used without further purification: cobalt acetate, dodecanoic acid, sodium borohydride are from Aldrich, iso-octane and hexane from Fluka, and sodium di(ethylhexyl) sulfosuccinate (NaAOT) from Sigma. The synthesis of cobalt(II) bis(2-ethylhexyl)sulfosuccinate, ($Co(AOT)_2$) was described previously [5]. Cobalt nanoparticles are synthesized by soft chemical reduction in reverse micelles under a N_2 atmosphere. The resulting spherical Co NCs, (7.4 nm in diameter) are coated with dodecanoic acid and has a 9.4 % size distribution (*i.e.* 68% of NCs have diameters equal to 7.4 nm ±0.7 nm). As synthesized, Co NCs possess highly disordered polycrystalline structure, with individual crystalline domains (typical size ≤ 1 nm) showing a likely fcc organization. The electronic diffraction pattern of as-synthesized Co NCs reveals (not shown here) two diffuse rings indicating a very low crystallinity. To form 3D supracrystals, a Highly Oriented Pyrolitic Graphite substrate (10 x 5 mm^2) is immersed horizontally in a colloidal solution of Co NCs dispersed in hexane. The solvent is then slowly evaporated at room temperature under N_2. The end product is 3D superlattices composed of a few hundred layers of Co NCs [6]. GISAXS study indicates that these supracrystals, mainly found at the border of the substrate, are characterized by long-range fcc mesoscopic order. The corresponding structural parameters: the stacking periodicity, the centre-to-centre interparticle distance (D_{c-c}) and the edge-to-edge interparticle distance (D_{i-p}) are found in Table 1 in the following section. After deposition, the 3D supracrystals are placed in a closed quartz capsule with a N_2 atmosphere and annealed at various temperatures (220, 250 and 350 °C) for 15 min. These samples are named; Native, S220, S250 and S350, respectively.

3. Results and Discussions

3.1. Nanocrystallinity and supracrystallinity evolution due to annealing

The annealing performed up to 220 °C does not impact the various mesoscopic organizations and thus structural parameters remain unchanged. The electronic diffraction pattern does not reveal any detectable nano-crystallographic transition compared to the native state. Conversely, annealing at 250 °C and higher induces significant improvement of the nanocrystallinity of NCs. In S250 sample, the diffraction pattern shows the coexistence of the (111) diffuse ring of likely fcc-Co and 5 diffuse rings (weaker in intensity) corresponding to 5 planes belonging to the Co hcp phase, indicating a clear structural transition from poorly crystallized fcc- to hcp-Co polycrystals having various coherence lengths. At the highest temperature (350 °C), the electronic diffraction and the HRTEM studies revealed predominantly hcp-single-crystalline NCs [7]. The GISAXS patterns shows that the fcc mesoscopic order of 3D supracrystals keep their integrity without coalescence at all annealing temperatures, although they become slightly *dense* with increasing temperature. This behavior, which is attributed to a decrease in D_{i-p} due to heating, is related to the supracrystal coherence length increase. The key supracrystalline structural parameter evolutions are listed in Table 1.

Table 1 : Supracrystal structural parameters

	Native	220 °C	250 °C	350 °C
(111) Stacking periodicity (nm)	8.49	8.49	8.32	8.00
Center-to-Center dist. D_{c-c} (nm)	10.4	10.4	10.2	9.8
Inter-particle distance, D_{i-p} (nm)	3	3	2.8	2.4

3.2. Magnetization Studies.

The nanocrystallinity effect was clearly seen in the magnetic behavior of supracrystals. As depicted in Figure 1 (left), the transition temperature T_c, defined as the position of zero-field-cooled (ZFC) magnetization peak, increases from 110 K (Native) to 290 K (S350) [7] passing through a minimum value of 102 K (S220). The ac susceptibility as a function of temperature and frequency showed usual 'critical slowing down' behavior signaling the existence of collective states below T_c in all samples (not shown). Using the Vogel-Fulcher model [8] for weakly-interacting systems, the particle anisotropy energy E_a and the effective temperature T_o (related to the interaction energy between NCs) are extracted. The anisotropy energy was found to increase from about 1360 to 1700 K, again exhibiting an anomalous behavior for the S220 (Figure 1 Right).

It has previously been suggested that at low annealing temperatures, where the re-crystallization is barely initiated, nanocrystalline structure become highly heterogeneous among NCs. The anomalous behavior as well as the broad ZFC-peak and a relatively high ZFC magnetization at the lowest temperature observed in S220 sample here are most likely due to such structural inhomogeneity among NCs [9]. Furthermore, the notion of 'average' values of E_a and T_o determined using the V-F law may no longer be valid in S220. As a general trend; however, we remark that the value of T_o, related to the dipolar interaction energy, grows much faster than the anisotropy energy; *i.e.*, E_a grows by mere 25% from Native to S350, whereas a nearly 400% gain was observed in T_o values between the two samples (from 53 K to 237 K). This can be partially explained **by** the reduced interparticle distance between NCs, the S350 sample in particular, that enhances the dipolar interactions among them (*cf.* Table 1). Thus the 'interaction-to-anisotropy' energy ratio becomes larger with improved nanocrystallinity, a condition favorable for the formation of a dipolar SFM state [3][10].

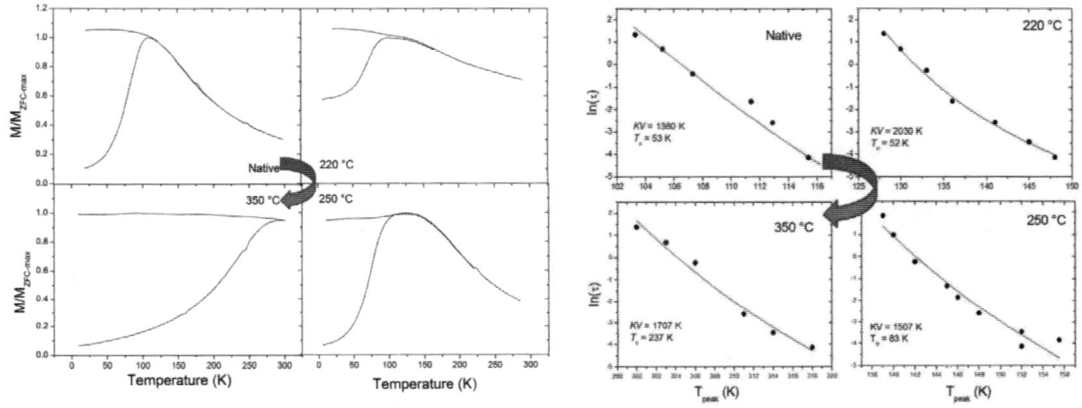

Figure 1: (left) ZFC/FC magnetization (normalized) measurements of Native, S220, S250 and S350 samples, presented clockwise from top left. All measurements were taken under an applied magnetic field of 10 Oe. (right) Evolution of frequency dependent transition (or collective blocking) temperature of all four samples. τ is the inverse of applied ac field frequency f ($\tau = 1/2\pi f$) and T_{peak} is the temperature at which in-phase ac-susceptibility attains the maximum value at a given frequency. Solid lines are fits to the V-F law.

4. Summary

Within long-range ordered 3D fcc supracrystals made of Co nanocyrstals, the nanocrystalline order combined with the mesoscopic order was found to influence both the NC anisotropy energy (E_a) as well as the interaction energy ($\sim T_o$) with T_o growing much faster than E_a. Further increase in interaction energy may result in the creation of a superferromagnetic state in 3D supracrystals.

References

[1] See for example, S. Singamaneni *et al. J. Mater. Chem.* **21** 16819 (2011).
[2] See for example, P. Nordblad, *J. Phys. D : App. Phys.* **41** 134011 (2008) and references therein.
[3] S. Bedanta *et al. J. Phys. D* **42** 013001 (2008).
[4] D. Parker *et al. J. Phys. Chem. Lett.* **1** 1139 (2010).
[5] I. Lisiecki *et al. Langmuir* **19**, 9486 (2003).
[6] I. Lisiecki *et al., Advanced Materials* **15** 712 (2003).
[7] D. Parker *et al. J. Phys. Chem. C* **111** 12632 (2007).
[8] See for example, J. L. Dorman *et al., Adv. Chem. Phys.* **98** 283 (1997) and references therein.
[9] Z. Yang *et al. J. Phys. Chem C* **116**15723 (2012)
[10] S. Mørup, *Hyperfine Interactions*, **90** 171 (1994); V. Russier, *J. App. Phys.* **89** 1287 (2001) ; M. Varon *et al., Sci. Rep.* **3** 1234 (2013).

Fe-oxide Nanoparticles: a natural playground for testing the ISP model

P Allia

Politecnico di Torino, DISAT, Corso Duca degli Abruzzi 24, 10129 Torino, Italy

E-mail: paolo.allia@polito.it

Abstract. Magnetic dipolar interaction plays a major role in systems of ferrimagnetic Fe-oxide nanoparticles. Dipolar interaction plays a role over a wide range of temperatures above blocking temperature and results in the onset of the Interacting Superparamagnetic (ISP) regime. In this paper, distinctive aspects and basic issues of the ISP model are discussed by exploiting the experimental data obtained in well-characterized magnetite nanoparticles.

1. Introduction

In recent years, Fe-oxide nanoparticles (NPs) have become one of the best study cases for testing models and concepts addressed to magnetic nanomaterials. The current chemical routes to ferromagnetic Fe-oxide NP production [1] have notable advantages, such as a quite reproducible output, a well-defined chemical composition and structure, a narrow spread of sizes/shapes. The synthesized NPs are often covered with a non-magnetic organic shell hindering NP aggregation by magnetic or electrostatic forces.

From a fundamentalist's viewpoint, magnetite NPs are an excellent playground for testing magnetic models. Experimental evidence gathered during the years points towards a substantial role played at all temperatures by interparticle interaction. In systems where magnetic contact interaction can safely be excluded, dipolar interaction plays a central role. At low temperatures, dipolar interaction is expected to bring about frozen collective states where the moment directions are dictated by the interplay of dipolar interactions rather than by intra-particle anisotropy. Frustrated moments and/or super-spin glass states are expected [2].

Systematic measurements indicate that the influence of dipolar interaction is not confined to the low-temperature domain, significant effects being measured even at much higher temperatures in systems of Fe-oxide NPs with a ferrimagnetic alignment of spins. These effects manifest themselves in the form of departures from both the SP magnetization law and the SP scaling; they are accounted for by the ISP model [3], basically a mean-field theory, which can be applied to temperatures where dipolar interaction energy, although no longer being dominant as at very low Ts, is nevertheless comparable in magnitude to the thermal energy k_BT. Indeed, the region where this condition is applicable extends over hundreds of Kelvin in actual NP systems.

The ISP-model prediction for the isothermal, anhysteretic magnetization of a monodisperse NP system containing N magnetic moments is $M_{ISP} = N\mu L[\mu H/k_B(T+T^*)]$ where L is the Langevin function and the term $k_BT^* = \alpha\mu^2/d^3 = \alpha N\mu^2 = \alpha M_s^2/N$ at the denominator of the Langevin function's argument depicts the dipolar energy, α being a number close to unity.

2. ISP vs. SP scaling law

In the ISP model the reduced magnetization of a monodisperse system is $m_{ISP} = L[\mu H/k_B(T+T^*)]$.

In the limit T >> T* the expression reduces to $L(\mu H/k_B T)$ and the standard SP scaling is obtained; however, when T < T*, the reduced magnetization is $L(\mu H/k_B T^*) \equiv L(H/\alpha M_s)$. Such a scaling law is actually observed in a number of Fe-oxide NP systems, as exemplified in Figure 1 referring to a typical magnetite nanopowder where the magnetic NPs, 10-12 nm in diameter, are covered by a substantial organic shell 2-3- nm thick (mostly acetylacetonate ligands).

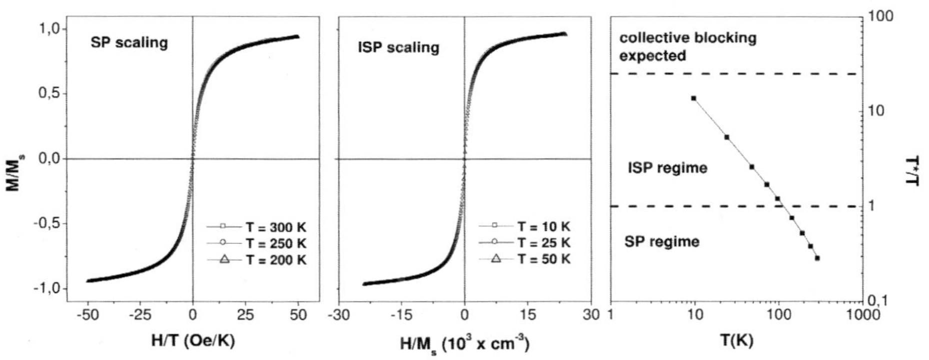

Figure 1 – SP and ISP scaling laws observed at high and low temperature in a magnetite nanopowder. The behavior of T*/T for this system is shown in the right panel.

3. Towards an explanation of the ISP model

Although the predictions of the ISP model are in good agreement with experimental data taken on well-characterized systems of Fe-oxide NPs, a difficulty arises about the form taken by the ISP expression. In fact, in the model the argument of the Langevin function of SP model is modified by changing the *denominator* (temperature) instead of the *numerator* (field), which seems to be an *ad-hoc* assumption.

The difficulty can be removed by considering that above T_B the magnetic moment directions are fluctuating in time by effect of thermal random torques. Re-orientation of magnetic moments in a SP material involves the crossing of a barrier whose main term is given by an intra-particle anisotropy energy. Starting from the pioneering works of Dormann et al., [4] various approaches have been developed in the past to extend the meaning and role of the barrier; suitably modified barriers have been introduced to account for inter-particle interactions as well.

The ISP model is based on the recognition that the *dynamical* aspects of the dipolar interaction in a NP system should not be neglected even when equilibrium/stationary properties are addressed. The randomly rotating i-th magnetic moment experiences a dipolar field H_{loc} which fluctuates in magnitude and direction – the associated Zeeman energy E_{DIP} fluctuates with the same rate and can be instantaneously positive or negative. The ensuing fluctuation of the barrier, which occurs at high rates at any T, adds to thermal effects and effectively contributes to *increase* the disorder of magnetic moments, in contrast to the ordering effect of an applied magnetic field. The typical fluctuation rate of dipolar field is $\tau_2^{-1} = \gamma H_{loc}$, γ being the gyromagnetic ratio; for magnetic NPs, $\tau_2^{-1} \cong 10^8 - 10^9$ Hz . Note that well above T_B the rate τ_2^{-1} is almost independent of T. The typical magnitude of the fluctuating field is $H_{loc} \sim \mu/d^3$ with d is a mean interparticle distance.

For a static barrier of height E_0 the escape rate of an individual moment from the local minimum can be reasonably thought of as given by a standard Arrhenius expression

$t^{1}_{esc} = \tau_0^{-1} \exp(-E_0/k_B T)$ where τ_0^{-1} is some pre-exponential factor of the order of 1×10^{10} Hz. In the ISP model, the denominator in the argument of the Langevin function is modified T to (T+T*), meaning that moments' dynamics takes place at equilibrium at an effective temperature higher than the actual temperature. This corresponds to assume that the escape rate of a re-ordering magnetic moment still follows an Arrhenius kinetics driven by a temperature higher than the actual one:

$$t^{1}_{esc} = \tau_0^{-1} \exp[-E_0/k_B(T+T^*)] \qquad (1)$$

It has been shown [5] that this assumption is compatible with first-principle approaches to generalized "particle" crossing of a fluctuating / flipping barrier done in the framework of the Langevin equation. In these approaches, a generalized "particle" in a bi-stable potential is submitted to a weak random driving force with a constant power spectral density (Gaussian white noise), so that the particle is able to cross from one minimum of the potential to the other. The height of the barrier separating the two minima of the potential is supposed to fluctuate in time. The solutions are typically numerical; however, a few of them can be cast in a simple form in particular cases (so-called "toy models") [6,7].

In NP systems barrier crossing involves *damped* rotation of magnetic moment μ, the barrier height being basically given in this case by the sum of a constant term related to the total magnetic anisotropy [4] plus a fluctuating term related to dipolar interaction. The expected barrier crossing rate induced by the barrier fluctuation rate τ_2^{-1} takes in this case a specific form [5] because in this case the generalized "particle" is the magnetization vector on a nanoparticle, and barrier crossing corresponds to damped coherent rotation of the magnetization across the overall barrier separating two adjacent minima of potential energy. In this case, t_{esc}^{-1} is written:

$$t^{-1}_{esc} = \frac{\gamma k_B T}{\lambda \mu \pi^2} f(\tau_2^{-1}, T) \qquad (2)$$

where λ is the damping constant of the Landau-Lifshitz-Gilbert's equation and $f(\tau_2^{-1},T)$ is a dimensionless analytic function whose explicit form depend on the particular model and whose behavior is given elsewhere [6-8]. This function indicates how much is barrier crossing enhanced by a fluctuation in barrier height, and is therefore strongly temperature-dependent (for instance, in the Bier-Astumian model [6], low-temperature barrier crossing is inhibited even in the presence of barrier height fluctuation). It should be noted that all models predict the same order of magnitude for t_{esc}^{-1} and τ_2^{-1} in magnetite NP systems around room temperature [5]. This substantiates the view that random magnetic moment rotation is enhanced by barrier fluctuation, particularly at high and intermediate temperatures, i.e., well above the blocking temperature.

For magnetite NPs, the following parameter values [5] are used: $\lambda \cong 0.1$-1, $\gamma \cong 1.5 \times 10^7$ Hz/Oe. The magnetic moment μ slightly decreases with temperature according to the experimental dependence of the measured saturation magnetization. The temperature dependence of t_{esc}^{-1} is therefore dominated by the function $f(\tau_2^{-1},T)$.

If the mean barrier's height is E_0 and a is the maximum amplitude of its fluctuation, so that the barrier can instantaneously take values between E_0-a and E_0+a, different analytical models apply depending on the value of the ratio between these quantities. When a < E_0 the results of the analytical formula proposed by Boguná et al. [8] for the Bier-Astumian model [6] are shown in Fig. 2(a) for a 10-nm magnetite system together with the expected temperature behavior of the barrier rate and the prediction of the ISP-modified Arrhenius plot (Eq.(1)). The static barrier case refers to the standard Arrhenius law. Instead, when the fluctuating term contribution is very large so that it becomes comparable to or even larger than E_0, the Döring-Gadoua [7] model applies. In fact, in Fe-oxide NP systems the fluctuating dipolar energy can be even larger than the static term. The analytical prediction for the study case where the barrier fluctuates between $\pm E_0$ is shown in Fig. 2(b) for two values of the LLG damping constants, together with the standard and ISP-modified kinetic predictions.

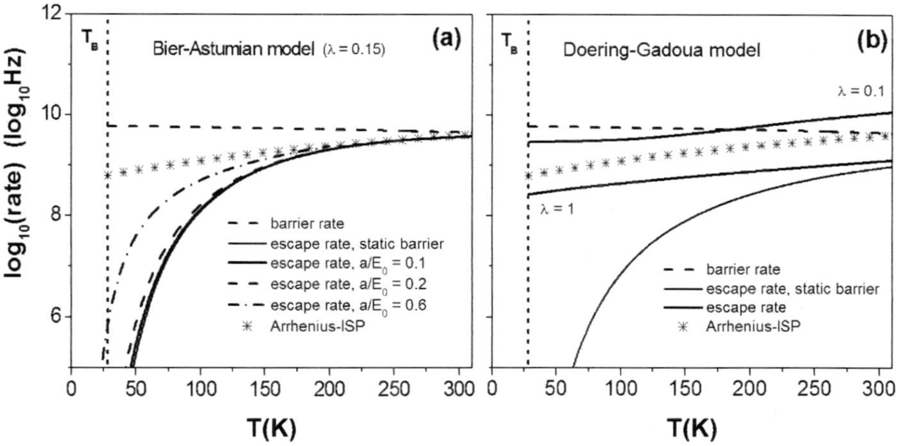

Figure 2 –(a): Comparison of the standard and ISP-modified Arrhenius kinetics with the predictions of the BA model of fluctuating-barrier particle crossing for different values of the fluctuating term; (b): the same for the DG model, for different values of the LLG damping constant. In both cases the expected barrier fluctuation rate $\tau_2^{-1} = \gamma H_{loc}$ is reported for comparison.

In both limiting cases, the ISP-modified Arrhenius law (Eq. 1) leads to a much weaker slowing down of the rotation rate for magnetic moments with decreasing temperature with respect to the standard Arrhenius prediction. Therefore, the ISP model matches the predictions of first-principle approaches for different values of the fluctuating barrier height over a wide interval of temperatures above T_B. The effect of a fluctuating barrier is basically to keep high the rate of random rotations of moments although if the temperature is decreasing; in turn, enhanced rotations of moments sustain the barrier's fluctuation in a positive feedback action; this effect is well represented by the ISP-modified Arrhenius kinetics.

In real systems the NP size distribution brings about a spectrum of values of both E_0 and H_{loc}.; consequently, the fluctuation rate of the dipolar field changes from nanoparticle to nanoparticle. The uniform dynamical parameters considered in this approach should therefore merely be regarded as average values of space-dependent quantities, which are nevertheless rather appropriate to take when a comparison with a mean-field theory is done. The present results substantiate the ISP model which is to be viewed as an approximation of a more complex, kinetic and statistical effect.

References

[1] Tartaj P, dal Puerto Morales M, Veintemillas-Verdaguer S, Gonzáles-Carreño T and Serna C J 2003 *J. Phys. D: Appl. Phys.* **36** R182
[2] Hiroi K, Komatsu K and Sato T 2011 *Phys. Rev.* **B83** 224423
[3] Allia P, Coisson M, Tiberto P, Vinai F, Knobel M, Novak M A and Nunes W C 2001 *Phys. Rev.* **B64** 144420
[4] Dormann J L, Fiorani D and Tronc E 1999 *J. Magn. Magn. Mater.* **202** 251
[5] Allia P and Tiberto P 2011 *J. Nanop. Res.* **13** 7277
[6] Bier and Astumian R D 1993 *Phys. Rev. Lett.* **7** 1649
[7] Döring C R and Gadoua J C 1992 *Phys. Rev. Lett.* **69** 2318
[8] Boguñá M, Porrà J M, Masoliver J and Lindenberg K 1998 1998 *Phys. Rev.* **E57** 3990

Measuring magnetic correlations in nanoparticle assemblies

M Beleggia[1,3] and C Frandsen[2]

1 Center for Electron Nanoscopy, Technical University of Denmark, Kgs. Lyngby, DK-2800 Denmark
2 Department of Physics, Technical University of Denmark, Kgs. Lyngby, DK-2800 Denmark

E-mail: marco.beleggia@cen.dtu.dk

Abstract. We illustrate how to extract correlations between magnetic moments in assemblies of nanoparticles from, e.g., electron holography data providing the combined knowledge of particle size distribution, inter-particle distances, and magnitude and orientation of each magnetic moment within a nanoparticle superstructure, We show, based on simulated data, how to build a radial/angular pair distribution function $f(r,\theta)$ encoding the spatial and angular difference between every pair of magnetic moments. A scatter-plot of $f(r,\theta)$ reveals the degree of structural and magnetic order present, and hence provides a measure of the strength and range of magnetic correlations.

[1] Introduction

Assemblies (or superstructures) of magnetic nanoparticles arranged over a superlattice represent a novel class of magnetic material where inter-particle exchange interactions may be negligible, and, hence, dipolar inter-particle interactions tend to dominate. As a result, magnetic superstructures behave differently than conventional magnets and their properties may be controlled and tuned by selecting the spacing and compositions of their constituents. Recently [1], we have studied dipolar interactions in self-assembled 15 nm cobalt particle structures using off-axis electron holography [2]. This technique enables the determination of orientation and magnitude of the magnetic moment of each nanoparticle [3] coupled with the structural ordering (particle arrangement and spacing) of the superlattice. Our study revealed that dipolar interactions are sufficiently strong to support long-range ferromagnetic order in quasi-2D superstructures, even when the lattice of nanoparticles is highly disordered. This observation supports the possibility of creating amorphous dipolar magnets, in contrast to the expectation that a disordered dipolar system may lead to spin-glass-like behaviour.

In order to quantify dipolar ferromagnetic order and reveal magnetic correlations, the first step is to extract the magnitude and orientation of each magnetic moment present in the structure. Considering that particles may be in the 10-20 nm size range, only a few techniques have the necessary single-particle spatial resolution and sensitivity to nanoscale magnetic fields. Electron holography [2] is one of such techniques: by creating an interference pattern between the electron wave scattered magnetically from the sample and a planar reference wave, it is possible to obtain a quantitative map of the projected local magnetic field within and around every nanoparticle in the system. The magnitude

3 To whom any correspondence should be addressed.

and orientation of each nanoparticle's (in-plane) magnetic moment is then measured from the local projected magnetic field, corrected with a proper demagnetization factor to accommodate the particles' shapes. Location, size and shape of each particle can be extracted e.g. from conventional bright field micrographs and particle-size analysis. Combining the information on particles arrangement with their overall magnetic state gives us access to the strength and range of magnetic correlations, as will be illustrated in the following section by the analysis of simulated data.

[2] The radial/angular pair distribution function

Consider a finite superstructure made of spherical single-domain nanoparticles with negligible magnetocrystalline anisotropy arranged over a planar close-packed lattice. Let the number of particles along the two axes be N and M, so that the total number of particles is $n=NM$; each particle, of radius R_i sits at location $\mathbf{r}_i$ and has a magnetic moment $\mathbf{m}_i$, with $i=1..n$. The quantities R_i, $\mathbf{r}_i$, $\mathbf{m}_i$ are assumed to be known; for example they can be measured from electron microscopy data as explained above, or by any other technique capable of providing the required information. We build the radial/angular distribution function $f(r,\theta)$ by calculating $r=|\mathbf{r}_i-\mathbf{r}_j|$ and $\theta=\arccos[\mathbf{m}_i\cdot\mathbf{m}_j/(m_im_j)]$ for every pair of particles. If the lattice is regular, r can only assume discrete values; with disorder, r varies continuously but will peak at every nearest neighbor distances, with peak widths proportional to the degree of disorder. We define the degree of lattice disorder σ as the ratio between the first neighbor peak half-width and the first neighbor distance.

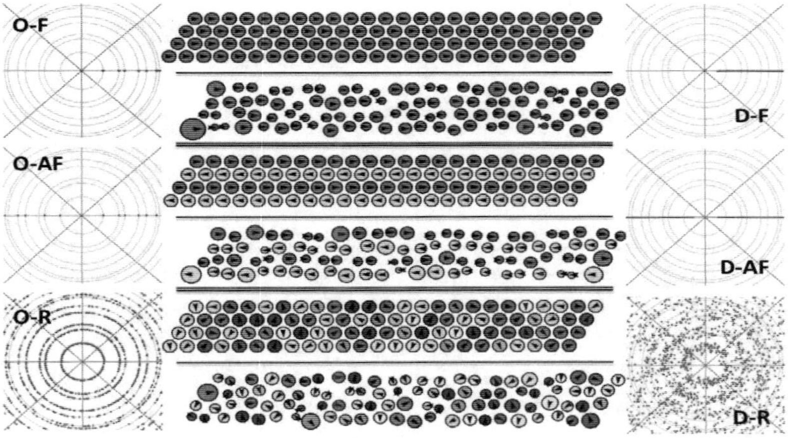

Figure 1. Radial/angular correlations (left, right) of ordered (O) and disordered (D) nanoparticle structures with imposed ferromagnetic (F), antiferromagnetic (A) and random (R) order.

The angular part reflects magnetic correlations: if the magnetic state is purely ferromagnetic, then $\theta=0$, independently of range; if the magnetic state is ferromagnetic only at short-ranges, then θ will be small for the first few nearest neighbor peaks, and then spread out at larger distances. To illustrate the behavior of $f(r,\theta)$ in a few representative scenarios, we consider a $N=24$, $M=4$ close-packed superstructure, in the outer shape of a sheared rectangle, structurally ordered (O, $\sigma=0$) and disordered

(D, σ=0.4), and impose three particular magnetic states upon it: ferromagnetic (F), antiferromagnetic (AF), random (R). As shown in Figure 1, the O-F system is characterized by clustering of the θ=0 (full alignment) points at the exact n^{th}-nearest neighbor distances (indicated with green circles), highlighting that magnetic correlations are present at long range; in contrast, the D-F system maintains the θ=0 independent on r, but the points are dispersed along the r-axis, broadening the possible neighbor distances distribution. The O-AF and D-AF systems behave similarly, except now we have two values of θ: θ=0, and $\theta=\pi$; note, however, the absence of the $\theta=\pi$ point on the 3rd-neighbor circle due to the particular configuration (no 3rd-neighbour pairs are antiparallel). The O-R shows clearly the clustering of θ points over the neighbor distance circles, with random spreading over the angle, indicating total loss of magnetic correlations; the D-R system shows θ points spread randomly through the angle and the distance, although some radial oscillations remain (barely) visible.

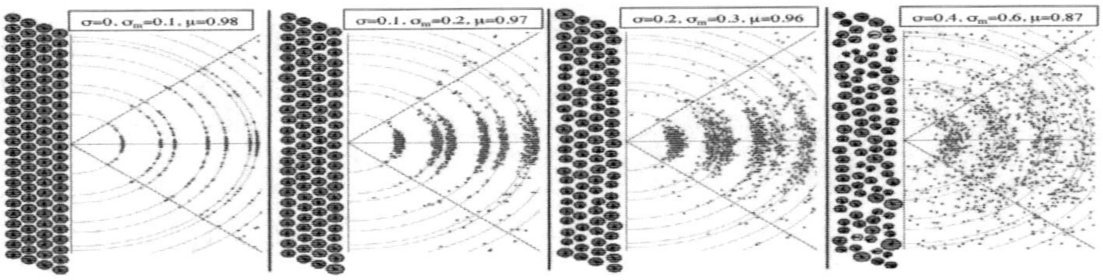

Figure 2. Superstructure of magnetic particles (N=24, M=4) with a mean diameter of 15 nm, a mean magnetic moment of 2.7 10^5 Bohr magnetons in a close-packed arrangement with lattice parameter of 20 nm and varying degree of structural disorder (σ=0,0.1,0.2,0.4 from left to right).

[3] Magnetic correlations as a function of lattice disorder

Allowing magnetic relaxation (at a temperature low enough to bypass superparamagnetic issues), the system evolves towards some energy minimum, reachable from the chosen initial condition. If we select relaxation from an initial F state, a situation representing the removal of a strong applied field and the relaxation to remanence, we can examine how magnetic correlations vary as a function of lattice disorder. Figure 2 shows four 24x4 systems, with σ=0,0.1,0.2,0.4; here we also measure the total net moment μ of the structure, normalized to its maximal value at the initial condition (saturation). The σ=0 case differs from the D-F system analyzed earlier in that now edge effects are playing a role introducing a number of high-θ points. As disorder increases, so does the angular spread of the magnetic moments, although even in the σ=0.4 case, where the lattice is barely distinguishable from amorphous, ferromagnetic correlations are still present: the angular difference near the nominal first neighbor distance has a standard deviation of about σ_m=0.6 rad=30°, meaning that at short-range most moments are aligned within 60°, and the assembly maintains a net remanent moment of 0.87.

In conclusion, we have illustrated how to measure magnetic correlations, and examined the radial/angular pair distribution function for 24x4 close-packed planar superstructures with varying disorder parameter, highlighting the role of dipolar interactions in maintaining long-range ferromagnetic order in quasi-2D assemblies of nanoparticles

References

[1] Varon M, Beleggia M, Kasama T, Harrison R J, Dunin-Borkowski R E, Puntes V F and Frandsen C 2013 *Sci. Rep.* **3**, 1234

[2] Tonomura A 1987 *Rev. Mod. Phys.* **59**, 639

[3] Beleggia M, Kasama T and Dunin-Borkowski R E 2010 *Ultramicroscopy* **110**, 425

8th International Conference on Fine Particle Magnetism (ICFPM2013) IOP Publishing
Journal of Physics: Conference Series **521** (2014) 012010 doi:10.1088/1742-6596/521/1/012010

Equilibrium and dynamic behaviour of (weakly) interacting assemblies of magnetic nanoparticles

F. Vernay,[1] Z. Sabsabi,[1] O. Iglesias[2] and H. Kachkachi[1]

[1]Laboratoire PROMES-CNRS UPR 8521, Université de Perpignan Via Domitia, Rambla de la thermodynamique - Tecnosud, 66100 Perpignan, France. [2]Dept. de Fisica Fonamental and Institut de Nanociencia i Nanotecnologia, Av. Diagonal 647, Universitat de Barcelona, Spain.

E-mail: `francois.vernay@univ-perp.fr`

Abstract. A still open issue related with the study of assemblies of magnetic nanoparticles, deposited on a substrate or embedded in a matrix, is that of the interplay between intrinsic features of the nanoparticles pertaining to their finite-size and boundary effects, and the collective effects entailed by their mutual interactions and their interactions with the hosting matrix or substrate. In this work we develop a semi-analytical approach that allows us to derive expressions for the magnetization and the susceptibility of interacting assemblies of single-domain ferromagnetic nanoparticles. We find that upon tuning the physical parameters pertaining to each nanoparticle or the shape of the assembly and its spatial arrangement, surface and inter-particle interactions may be set up to play additive or comptetive roles leading to assemblies with optimal magnetic properties.

1. Introduction

Over the last two decades, the development of practical applications for magnetic nanoparticle assemblies has raised several fundamental questions such as how to take account of the role of surface effects in the magnetic properties of an individual nanoparticle? What is the role of inter-particle interactions in nanoparticle assemblies? If these two questions have been separatly answered many years ago, for instance both at the numerical or analytical levels concerning the surface effects [1, 2, 3] or the dipole-dipole interactions (DDI) [4, 5, 6, 7, 8], only recent investigations on static magnetic properties take a systematic approach where intrinsic and collective terms contributing to the energy are taken explicitly into account [9, 10]. Tackling the problem of the interplay between surface effects and DDI can in principle be performed in different ways, yet as a mutli-spin formulation for the individual particles is tractable only numerically, the present work relies on an effective description of macrospins where the surface effect induces a cubic anisotropy as it has been shown in Ref.[2]. Furthermore, we restrict our study to the case of monodisperse weakly interacting assemblies where the uniaxial anisotropies are aligned with the external field. These simplifications enable us to derive semi-analytical expressions in which the parameters controlling the surface anisotropy and the DDI clearly appear. The paper is organized as follows: In the first section we present our model and briefly present the calculations leading to the final expressions of the magnetization and the AC susceptibility, a second section is devoted to the presentation of our results. The paper ends with our concluding remarks.

2. Model: energy, magnetization and susceptibility

We consider an assembly of $\mathcal{N}$ ferromagnetic nanoparticles each carrying a magnetic moment $\boldsymbol{m}_i = m_i \boldsymbol{s}_i$, $i = 1, \cdots, \mathcal{N}$ of magnitude m and direction $\boldsymbol{s}_i$, with $|\boldsymbol{s}_i| = 1$. The nanoparticles are distributed on a simple cubic lattice of parameter a. Each magnetic moment has a uniaxial easy axis $\boldsymbol{e}$ aligned in the same z direction. The energy of a magnetic moment $\boldsymbol{m}_i$ interacting with the whole assembly, in a magnetic field $\mu_0 \boldsymbol{H} = H\boldsymbol{e}_z$ (so that H is measured in Tesla), reads (after multiplying by $-\beta = -1/k_B T$)

$$\mathcal{E}_i = \mathcal{E}_i^{(0)} + \mathcal{E}_i^{\mathrm{DDI}}, \quad \text{where} \quad \mathcal{E}_i^{(0)} = x\boldsymbol{s}_i \cdot \boldsymbol{e}_z + \sigma \left[(\boldsymbol{s}_i \cdot \boldsymbol{e}_z)^2 - \frac{\zeta}{2} \left(s_{i,x}^4 + s_{i,y}^4 + s_{i,z}^4 \right) \right] \tag{1}$$

is the energy of the free nanoparticle at site i with the first term being the Zeeman energy. The second term is the anisotropy energy comprising contributions from the core and the surface, such that the model is an effective one spin problem (EOSP) with the assumption that $|\zeta| < 1$. This formulation with both a uniaxial and a cubic effective anisotropies has been shown to model the effect of surface anisotropy on the global magnetization of a nanoparticle [2]. Next, $\mathcal{E}_i^{\mathrm{DDI}}$ is the DDI energy defined by $\mathcal{E}_i^{\mathrm{DDI}} = \xi \sum_{j<i} \left[3(\boldsymbol{s}_i \cdot \boldsymbol{e}_{ij})(\boldsymbol{e}_{ij} \cdot \boldsymbol{s}_j) - \boldsymbol{s}_i \cdot \boldsymbol{s}_j \right] / r_{ij}^3$ where $\boldsymbol{r}_{ij} = \boldsymbol{r}_i - \boldsymbol{r}_j$, $\boldsymbol{e}_{ij} = \boldsymbol{r}_{ij}/r_{ij}$. We make here the assumption that the dipole-dipole approximation is valid for the considered samples, *i.e.* spherical nanoparticles with a reasonable interparticle distance such that the inner structure of the particles can be neglected. We have introduced the following dimensionless parameters

$$x \equiv \frac{mH}{k_B T}, \quad \sigma \equiv \frac{K_2 V}{k_B T}, \quad \zeta \equiv \frac{K_4}{K_2}, \quad \xi = \left(\frac{\mu_0}{4\pi} \right) \left(\frac{m^2/a^3}{k_B T} \right),$$

together with the DDI coefficient $\tilde{\xi} \equiv \xi \mathcal{C}^{(0,0)}$, where $\mathcal{C}^{(0,0)}$ is a lattice sum and for a cubic sample $\mathcal{C}^{(0,0)} = -4\pi \left(D_z - \frac{1}{3} \right)$, D_z being the demagnetizing factor along z [11]. To make contact with experiments, the applied field H can be estimated: at $T = 5K$ for $x = 5$ in the case of a 3nm cobalt nanoparticle (*i.e.* $m \approx 2167\mu_B$) one finds $H \approx 17$mT. In this case, the present approach remains valid in the limit of weak DDI, a critical value of the volume concentration can be estimated $C_{\mathrm{v_c}} = \frac{\pi^2 k_B T}{\mu_0 m^2} \frac{D^3}{\sqrt{2\mathcal{R}}}$,[1] which corresponds to a critical interparticle distance $a_c \approx 4D = 12$ nm.

With the definition of the energy given above and assuming that the considered assemblies are diluted ($\tilde{\xi} \ll 1$), the dipolar term $\mathcal{E}_i^{\mathrm{DDI}}$ in Eq. (1) is considered as a perturbation to the free particle energy $\mathcal{E}_i^{(0)}$. An expansion in x can be performed for low fields within this EOSP formulation in the presence of weak DDI. After some algebra the magnetization can be computed, with the corrections linear in $\tilde{\xi}$ stemming from the interactions, and can be expressed as $m\left(x, \sigma, \zeta, \tilde{\xi}\right) \simeq \chi^{\mathrm{eq}}\left(x, \sigma, \zeta, \tilde{\xi}\right) x$, where $\chi^{\mathrm{eq}}\left(x, \sigma, \zeta, \tilde{\xi}\right) \simeq \chi_{\mathrm{free}}^{\mathrm{eq}} + \tilde{\xi}\chi_{\mathrm{int}}^{\mathrm{eq}}$, with $\chi_{\mathrm{free}}^{\mathrm{eq}}$ being the equilibrium (linear) susceptibility of the noninteracting assembly in the limit of high anisotropy energy barrier [10, 11]

$$\chi_{\mathrm{free}}^{\mathrm{eq}}(x, \sigma, \zeta) = 2\chi_0^{\perp}\sigma \left[\chi_{\mathrm{free}}^{(1)} + 3\chi_{\mathrm{free}}^{(3)} x^2 \right], \tag{2}$$

$$\chi_{\mathrm{free}}^{(1)} = \left(1 - \frac{1}{\sigma} \right) + \frac{\zeta}{\sigma} \left(-1 + \frac{2}{\sigma} \right) \quad , \quad \chi_{\mathrm{free}}^{(3)} = \frac{1}{3} \left[\left(-1 + \frac{2}{\sigma} \right) + \frac{\zeta}{\sigma} \left(2 - \frac{5}{\sigma} \right) \right].$$

Here $\chi_0^{\perp}$ is the transverse equilibrium susceptibility per spin at zero temperature in the absence

[1] $\mathcal{R}$ being a lattice sum introduced in Ref. [11]

of a bias field $\chi_0^{\perp} \equiv \left(\frac{\mu_0 m^2}{2K_2 V}\right)$. The interaction contribution to the susceptibility reads [10]

$$\chi_{\rm int}^{\rm eq}(x,\sigma,\zeta) = 2\chi_0^{\perp}\sigma\left[\chi_{\rm int}^{(1)} + 3\chi_{\rm int}^{(3)}x^2\right], \quad (3)$$

$$\chi_{\rm int}^{(1)} = 1 - \frac{1}{\sigma} - 2\left(1 - \frac{3}{\sigma}\right)\frac{\zeta}{\sigma} \quad , \quad \chi_{\rm int}^{(3)} = -\frac{4}{3}\left[\left(1 - \frac{3}{\sigma}\right) - \frac{3\zeta}{\sigma}\right].$$

The dynamical properties of a nanoparticle assembly can be studied, for instance, thanks to the AC susceptibility. In the high-energy barrier limit and for low frequencies, assuming a single dynamical mode characterized by a relaxation rate Γ, the AC susceptibility can be derived within the Debye formalism $\chi(\omega) = \frac{\chi_{\rm eq}}{1+i\omega\Gamma^{-1}}$. It has been shown [12, 13] that the over barrier relaxation rate of a weakly interacting assembly can be expressed in terms of the relaxation rate of a free particle Γ_0 and the longitudinal and transverse components of the dipolar field ($\Xi_\parallel$ and $\Xi_\perp$), *i.e.* $\Gamma \simeq \Gamma_0\left[1 + \frac{1}{2}\left\langle \Xi_\parallel^2\right\rangle_0 + \frac{1}{4}F(\alpha)\left\langle \Xi_\perp^2\right\rangle_0\right]$. The dynamics within the potential well is taken care of with the help of the function $F(\alpha)$ which depends on the damping parameter α. In the limit of intermediate to high damping, Γ_0 can be derived within Langer's formalism [14]: care has to be taken while deriving the expression of Γ_0 in the EOSP model, as the energy landscape is modified by the presence of a cubic anisotropy controlled by the parameter ζ.

3. Magnetization curves

Our results are summarized in Fig. 1 where we give two examples of magnetization curves for two assemblies of $\mathcal{N} = 2000$ particles with different shapes: oblate ($20 \times 20 \times 5$) and prolate ($10 \times 10 \times 20$). More results can be found in Ref. [10]. In order to see the effect of the DDI we have chosen samples with different volume concentrations $C_{\rm v} \propto \xi$: either $C_{\rm v} \sim 0$ or $C_{\rm v} = 0.08\%$. We choose here to keep the intrinsic parameter ζ (reflecting the surface effect) positive. In this case, the easy axes of the cubic anistropy term in Eq. (1) lie along the main diagonals of the cube, which means that the surface effect tends to reduce the magnetization. This intrinsic effect is clearly seen in both plots if one compares the black (non-interacting assemblies without surface effect) and red curves (non-interacting assemblies with $\zeta = 0.25$). In contrast, the role of DDI

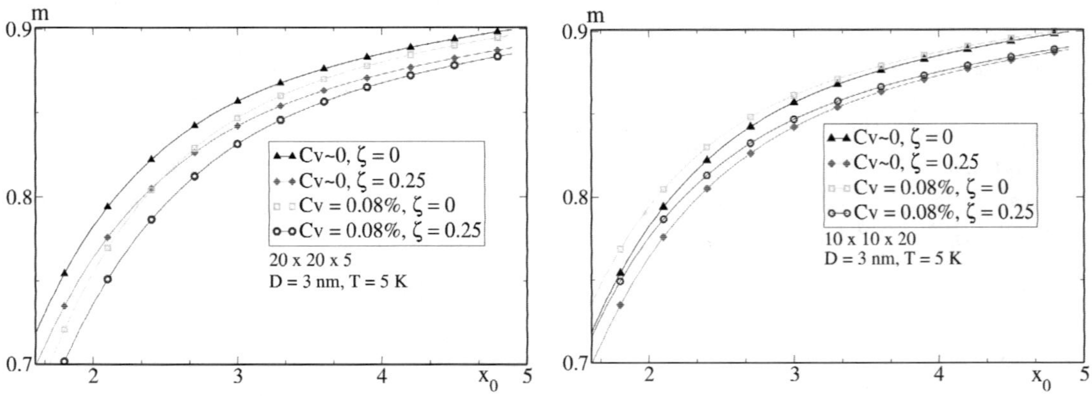

Figure 1. Magnetization as a function of the field x for monodisperse assemblies of particles with diameter D = 3 nm at $T = 5K$ for an oblate ($20 \times 20 \times 5$) with $\mathcal{C}^{(0,0)} \approx -4.086$, or a prolate ($10 \times 10 \times 20$) sample with $\mathcal{C}^{(0,0)} \approx 1.73$. The volume concentration $C_{\rm v} = 0.08\%$ corresponds to an interparticle distance $a \approx 8.68D$.

strongly depends on the sample's shape: whilst the DDI effective field induces an easy xy plane for the oblate sample, it facilitates magnetization along the z axis for the prolate sample. This results in a reduction of the magnetization in one case, green curve for the oblate sample, and an enhancement of the magnetization for the prolate one. Hence, for a concentration $C_v = 0.08\%$ at finite parameter $\zeta = 0.25$ controlling the surface effects, we see that the collective (DDI) and intrinsic effects can either be additive (oblate) or can mutually screen each other (prolate). This is directly in line with the low field expression of the magnetization obtained in the previous section. Indeed, the expression of the magnetization depends on the product $\zeta\tilde{\xi}$. In the present case, the sign of $\tilde{\xi}$ changes because the lattice sum $\mathcal{C}^{(0,0)}$ goes from negative to positive values as the sample's shape changes from oblate to prolate.

4. Concluding remarks

The EOSP approach adopted here in order to investigate the interplay between surface and DDI effects in ferromagnetic nanoparticles assemblies enabled us to derive semi-analytical expressions for the magnetization. We presented the low-field expansion of the latter which is directly related to the equilibrium susceptibility. We clearly see that the magnetic properties of the whole assembly can be tuned by the product $\zeta\xi\mathcal{C}^{(0,0)}$ where ζ controls the effects related to the surface and $\mathcal{C}^{(0,0)}$ depends on the sample's shape as we have seen in the specific example highlighted here. Furthermore, by modifying the intrinsic physical parameters related with the size, shape and underlying crystallographic structure the sign and amplitude of ζ are modified. Hence, it is relatively straightforward within this approach to infer whether DDI and surface effects are in competition or not. In addition, as we have suggested, the present work sets the stage for further investigations of the dynamical properties [14].

Acknowledgements

O. I. acknowledges funding from the Spanish MINECO (MAT2012-33037), Catalan DURSI (2009SGR856) and European Union FEDER funds (Una manera de hacer Europa).

References

[1] RH Victora and JM McLaren 1993 *Phys. Rev. B* **47** 11583
[2] DA Garanin and H Kachkachi 2003 *Phys. Rev. Lett.* **90** 65504
[3] P-M Déjardin, H Kachkachi, Yu Kalmykov 2008 *J. Phys. D* **41** 134004
[4] J-L Dormann, D Fiorani, and E Tronc 1997 *Adv. Chem. Phys.* **98** 283
[5] P Allia, M Coisson, P Tiberto, F Vinai, M Knobel, M A Novak, and W C Nunes 2001 *Phys. Rev. B* **64**(14) 144420
[6] Beleggia M, Tandon S, Zhu Y and de Graef M 2004 *Journal of Magnetism and Magnetic Materials* **278** 270–284
[7] H Kachkachi and M Azeggagh 2005 *Eur. Phys. J. B* **44** 299
[8] D Ledue and R Patte and H Kachkachi 2012 *J. Nanoscience and Nanotechnology* **12** 4953
[9] G Margaris, KN Trohidou, and H Kachkachi 2012 *Phys. Rev. B* **85** 024419
[10] Z Sabsabi, F Vernay, O Iglesias, H Kachkachi 2013 *Phys. Rev. B* **88** 104424
[11] PE Jonsson and JL Garcia-Palacios 2001 *Phys. Rev. B* **64** 174416
[12] JL Garcia-Palacios 2000 *Adv. Chem. Phys.* **112** 1
[13] PE Jonsson and JL Garcia-Palacios 2001 *Europhys. Lett.* **55** 418
[14] F Vernay, Z Sabsabi, H Kachkachi, in preparation 2014

Journal of Physics: Conference Series **521** (2014) 012011 doi:10.1088/1742-6596/521/1/012011

Ideal superspin glass behaviour in a random-close-packed ensemble of maghemite nanoparticles

J A De Toro[1], S S Lee[2], R Mathieu[3], P S Normile[1], D Salazar[1], J L Cheong[2], P Muñiz[1], J M Riveiro[1], M Hillenkamp[4], A Tamion[4], F Tournus[4], P Nordblad[3]

[1]Instituto Regional de Investigación Científica Aplicada (IRICA) and Departamento de Física Aplicada, Universidad de Castilla-La Mancha, 13071 Ciudad Real, Spain

[2]Institute of Bioengineering and Nanotechnology, 31 Biopolis Way, The Nanos, Singapore 138669, Singapore

[3]Department of Engineering Sciences, Uppsala University, Box 534, SE-751 21 Uppsala, Sweden

[4]Institut Lumière Matière, UMR5306 University Lyon 1-CNRS, 69622 Villeurbanne, France

E-mail: joseangel.toro@uclm.es

Abstract. Highly uniform (8 nm) *bare* maghemite nanoparticles were pressed into a disc with a volume fraction close to the characteristic filling factor of random-close-packed ensembles of spherical particles. We review the ideal superspin glass behaviour exhibited by this material, including an onset of the absorption component of the ac susceptibility at the freezing transition as sharp (in a normalized temperature scale) as those observed in *atomic* spin glasses, narrow memory dips in the zero-field cooled magnetization, and a spin-glass characteristic field-dependence of the susceptibility. Critical exponents were extracted from static and dynamic scaling analysis.

1. Introduction
Increasing possibilities in the control of nanoparticle (NP) uniformity and spatial arrangement have prompted magnetic studies on super- (or supra-) crystals, ordered 3D arrays of NPs [1]. However, their "super-amorphous" counterparts, the *a priori* simpler random-close-packed (RCP) ensembles of NPs, have been scarcely approached. Here we demonstrate *ideal* superspin glass (SSG) behaviour in one such system of *bare* ferrimagnetic oxide NPs. Compared to previous reports on strongly interacting but less concentrated frozen ferrofluids, granular films and "packed" powders of typically ill-defined filling factor (see references in Ref. 2), the properties of this material are found to be remarkably close to those of conventional *atomic* spin glasses.

2. Experimental
Highly crystalline, very uniform *bare* maghemite (γ-Fe_2O_3) NPs were obtained after the complete removal of the oleic acid surfactant (employed in their synthesis) by thermal decomposition [3]. These

bare NPs were compacted under a uniaxial pressure of 0.7 GPa to form a disc. The particles were characterized by transmission electron microscopy (TEM), as well as small (SAXS) and wide (WAXS) angle X-ray scattering. The pressed disc density was measured by the Archimedes'method. A Quantum Design MPMS magnetometer was employed to characterize the field and temperature dependence of the ac and dc susceptibility after different standard protocols (zero-field-, ZFC, and field-cooling, FC, as well as ZFC protocols with aging stops during the cooling).

3. Results and discussion

The TEM micrograph in Figure 1 shows the high uniformity of the *bare* nanoparticles (2% diameter dispersion), with an average diameter of 8.0 nm, employed to prepare the pressed disc displayed in the inset. The oscillations for Q (scattering vector) > 1 nm^{-1} in the SAXS pattern recorded in a loose powder of oleic-acid coated NPs (right panel in Figure 1) consistently indicate a well-defined NP form factor and, thus, particle diameter, whereas the first peak relates to the random configuration of the particles [3]. The filling factor of the disc (67 %), derived directly from its density and that of the constituent NPs (measured by XRD), is only slightly higher than that of an array of random-close-packed hard spheres (64 %), justifying the name "RCP" for this sample.

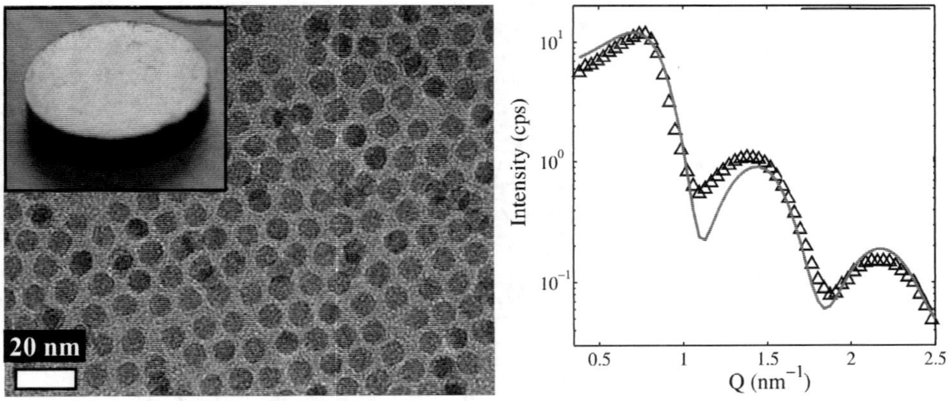

Figure 1. Left: transmission electron micrographs of bare maghemite nanoparticles and a photograph of the random-close-packed sample (RCP), a 6 mm diameter disc prepared with the bare particles. Right: SAXS from oleic-acid coated NPs, fitted as described in Ref. 3. Adapted from Ref. 3.

The RCP sample exhibits FC and ZFC magnetization curves (Figure 2) typical of spin glasses, but with a ZFC peak temperature T_{peak} = 145 K about four times (an unusually high factor) the blocking temperature of the NPs (T_B = 36 K) as measured in a dilute reference sample prepared with the same particles coated with thick (17 nm) silica shells [3]. This indicates the presence of particularly strong dipolar interactions in the RCP sample.

The presence of the characteristic chaotic dynamics of the spin-glass state is unequivocally manifested by the so-called "memory effect" [4], also plotted in Fig. 2 (left panel): when the system is zero-field cooled, stops (of 4 hours each) performed at some temperatures below the freezing temperature ($T_g \approx T_{peak}$), in this case T_{S1} = 110 K and T_{S2} = 70 K, during the cooling produce dips at the stop temperatures when the magnetization is subsequently recorded upon heating from the base temperature. The memory curve coincides with the reference curve at low temperatures (i.e. the system *rejuvenates*) and it merges with the reference curve at temperatures sufficiently far from the stop temperatures. The very faint memory registered at the lowest stop temperature T_{S3} = 30 K (< T_B in the dilute reference sample) reflects the fact that most particles are blocked on the timescale of the experiment, and therefore the system is not observed to age. Both memory dips in Fig. 2 exhibit a 5%

depth and a width at half maximum of $\Delta T \approx 17$ K or, in relative units, $\Delta T/T_g \approx 12\%$. This parameter is smaller than those previously measured in a variety of non-close-packed superspin glasses (see references in Ref. 2). Atomic spin-glasses exhibit still somewhat narrower dips than the *RCP* sample [5].

The magnetic dynamics of the sample was studied by ac susceptibility. The freezing transition is signalled by a cusp in the temperature dependence of the in-phase component χ' (see Ref. 2) and a sudden onset in the out-of-phase component χ'' [2]. The right panel in Figure 2 compares this unusually abrupt onset of the absorption component with data from previous reports of superspin glasses as well as from a model atomic spin glass. Not only the RCP sample improves the sharpness of the transition observed in previous quality particle systems, it also becomes remarkably close to the behaviour observed in model *atomic* spin glasses.

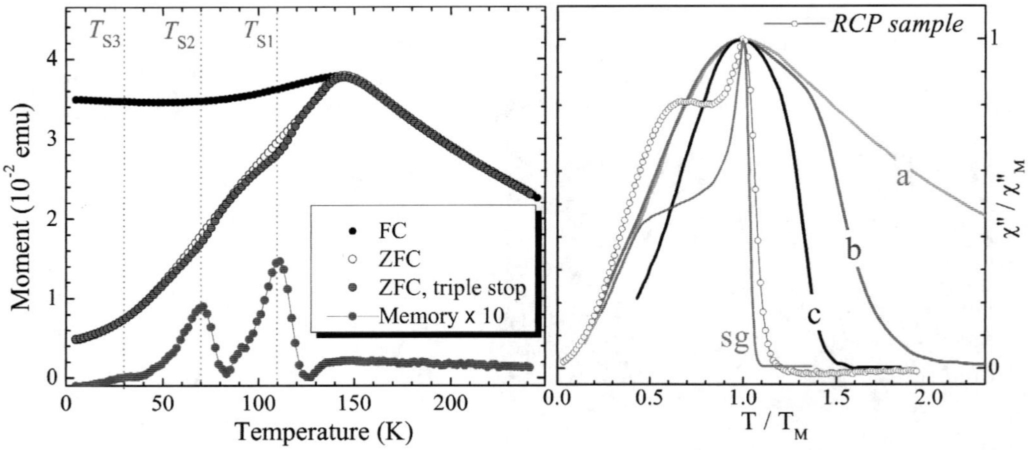

Figure 2. Left: thermal dependence of the magnetization (measured in an applied field of 5 Oe) and memory effect after a triple-stop (of 4 hours each at the indicated temperatures) during cooling. Right: normalized comparison of the absorption component in the RCP system with previous reports of SSGs (a = Ref. 6, b = Ref. 7, c = Ref. 8) and with a model *atomic* spin glass (sg = Ref. 9). Adapted from Ref. 2.

The critical character of the glass transition was further tested with both dynamic and static scaling analyses, and the corresponding critical exponents were extracted (see Refs. 2 and 10). The peak in $\chi'(T)$, as well as the ZFC maximum in Figure 2, reflect the system's optimum response to the driving field when its characteristic relaxation time, τ, equals the experiment observation time $t_m=1/\omega$. Thus, the temperature dependence of this relaxation time can be found from the shift of the $\chi'(T)$ peak temperature with increasing frequency ω of the driving ac field. This $\tau(T)$ data was successfully fitted to the critical slowing down power law $\tau = \tau_* \varepsilon^{-z\nu}$, where $\varepsilon = \left(T_p/T_g - 1\right)$ is the reduced temperature and the prefactor $\tau_* = \tau_0 \exp\left(KV/k_B T\right)$ is the temperature dependent Néel-Brown relaxation time of the individual particles (which does not introduce any new fitting parameters, since the NP anisotropy barrier KV and attempt time τ_0 were previously determined using the above mentioned reference dilute sample [2]), yielding T_g = 140.1(4) K and a critical exponent zv = 9.5(2), in good agreement with values of zv reported in long-range SGs as well as in some SSGs [8].

The ac susceptibility was found to be very sensitive to superposed dc-fields, as is usually observed in superspin glasses [10]. Recording the ac-susceptibility as a function of dc magnetic field above T_g can be used to extract the non-linear susceptibility of the system and check whether the system undergoes a true spin-glass phase transition. The magnetization M in a spin glass can be expressed as $M = \chi_1 H + \chi_3 H^3 + \chi_5 H^5 + ...$, where χ_1 is the linear susceptibility and the higher order terms represent the non-linear susceptibility defined as $\chi_{nl}(H) = \chi_1 - M/H$. In a spin glass all the non-linear coefficients, like, e.g., χ_3, diverge at T_g [10]. The non-linear susceptibility can be conveniently extracted from ac susceptibility data as a function of superposed dc fields, since above T_g (where $\chi'' = 0$ and, thus, $\chi = \chi'$ and $\chi_{nl} = \chi'_{nl}$) the dc field dependence of the nonlinear susceptibility is given by $\chi_{nl}(H) = \chi'_{nl}(H) = \chi'_1 - M/H = \chi'(H=0) - \chi'(H)$. It can be shown that for small dc fields (but still much larger than the ac driving field) $\chi_{nl} \propto -3\chi_3 H^2$ [10]. Fitting the low field region of the so-obtained $\chi_{nl}(H;T)$ data to such expression, we can extract χ_3, the lowest order term of χ_{nl}, at each of the measured temperatures (see Ref. 10). At any phase transition $\chi_3(T)$ is expected to diverge with the reduced temperature $\varepsilon = (T - T_g)/T_g$ as $\chi_3 \propto \varepsilon^{-\gamma}$ [10,11]. This is the case indeed for the data extracted for the RCP sample, where $\chi_3(T)$ indeed diverges at $T_g = 140$ K with an associated critical exponent $\gamma = 2.5(5)$. This confirms the occurrence of the spin glass phase transition suggested by the dynamical scaling analysis described above. Further scaling analysis of the complete $\chi_{nl}(T,H)$ data suggests a magnetization exponent $\beta = 0.2(1)$ [10,11]. Within the uncertainties observed in the literature, the values obtained for γ and β appear closer to those associated with atomic spin glasses than to those reported for superspin glasses.

Finally, it has been shown that interparticle superexchange interactions between the NPs in the RCP sample play a minor role compared to classical dipolar interactions in establishing the collective, superspin-glass state (see Ref. 3). This result was concluded in a comparison of the ZFC peak temperatures (and aging behaviour) in a series of five samples comprising identical maghemite particles in different concentrations, where the concentration was accurately controlled by means of silica shells of different thickness. We found that the freezing temperature of the most concentrated samples in the series (those with $0 \leq$ shell thickness < 3 nm), where the ZFC memory effect was also observed, vary in direct proportionality with the volume fraction of the maghemite cores and therefore with the strength of dipolar interactions [3].

Summarizing, we have prepared and characterized what could be considered a benchmark superspin glass, where the optimization of NP uniformity, spatial randomness and strength of dipolar interactions (interparticle exchange has been proven to be negligible) leads to archetypal spin glass behavior.

Acknowledgments

Financial support from the Spanish CICYT [MAT2011-26207] and the Swedish VR.

References

[1] O. Kasyutich, R. D. Desautels, B. W. Southern, and J. van Lierop 2010 *Phys. Rev. Lett.* **104**, 127205.
[2] J. A. De Toro, S. S. Lee, D. Salazar, J. L. Cheong, P. S. Normile, P. Muñiz, J. M. Riveiro, M. Hillenkamp, F. Tournus, A. Tamion, and P. Nordblad 2013 *Appl. Phys. Lett.* **102**, 183104.
[3] J. A. De Toro, P. S. Normile, S. S. Lee, D. Salazar, J. L. Cheong, P. Muñiz, J. M. Riveiro, M. Hillenkamp, F. Tournus, A. Tamion, and P. Nordblad 2013 *J. Phys. Chem. C* **117**, 10213.
[4] R. Mathieu, P. E. Jönsson, D. Nam, and P. Nordblad 2001 *Phys. Rev. B* **63**, 092401.
[5] P. E. Jönsson, H. Yoshino, H. Mamiya, and H. Takayama 2005 *Phys. Rev. B* **71**, 104404.
[6] M. Suzuki, S. Fullem, I. Suzuki, L. Wang, and C.-J. Zhong 2009 *Phys. Rev. B* **79**, 024418.
[7] D. Parker, V. Dupuis, F. Ladieu, J. Bouchaud, E. Dubois, R. Perzynski, and E. Vincent 2008 *Phys. Rev. B* **77**, 104428.
[8] M. F. Hansen, P. E. Jönsson, P. Nordblad, and P. Svedlindh 2002 *J. Phys.: Condens. Matter* **14**, 4901–4914.
[9] J. Mattsson, T. Jonsson, P. Nordblad, H. Aruga Katori, and A. Ito 1995 *Phys. Rev. Lett.* **74**, 4305.
[10] R. Mathieu, J. A. De Toro, D. Salazar, S. S. Lee, and P. Nordblad 2013 *EPL* **102**, 67002.
[11] R. Mathieu and Y. Tokura 2007 *J. Phys. Soc. Jpn.* **76**, 124706.

Super spin dimensionality of a mono-dispersed and densely packed magnetic nanoparticle system

M S Andersson[1], J A De Toro[2], S S Lee[3], R Mathieu[1] and P Nordblad[1]

[1]Department of Engineering Sciences, Uppsala University, Box 534, SE-751 21 Uppsala, Sweden

[2]Instituto Regional de Investigación Científica Aplicada (IRICA) and Departamento de Física Aplicada, Universidad de Castilla-La Mancha 13071 Ciudad Real, Spain

[3]Institute of Bioengineering and Nanotechnology, 31 Biopolis Way, The Nanos, Singapore 138669, Singapore

E-mail: `mikael.andersson@angstrom.uu.se`

Abstract. The dynamics of a dense near mono-dispersed assembly of maghemite nanoparticles is investigated by measurements of the temperature dependence of the isothermal remnant magnetization induced by temporal application of weak magnetic fields at constant temperature. The results suggest that the dimensionality of the super spins of the particles is of Heisenberg character at high temperatures but crossover to become Ising like at lower temperatures.

1. Introduction

It has been demonstrated that dense magnetic nanoparticle systems attains spin glass like properties at low temperatures. [1, 2] However, the existence of a super spin glass temperature and conventional critical slowing down has only been evidenced in systems with narrow distributions of the particle sizes and provided that the Arrhenius behavior of the individual particle relaxation time is accounted for. [3, 4, 5]

Certain dynamic properties of atomic spin glasses have been found to depend on the spin dimensionality of the system, e.g. the temperature dependence of the weak field remnant magnetization attained in an isothermal protocol [6]. Here we employ this measurement protocol to investigate whether a densely packed system of near mono-dispersed maghemite nanoparticles [7] can be classified as consisting of Ising or Heisenberg super spins.

2. Experimental

The experiments were made in a Quantum Design MPMS SQUID magnetometer. Zero field cooled (ZFC) and field cooled (FC) magnetization were recorded as a function of temperature to characterize the sample and to ensure that weak enough fields were used to achieve linear response of the ZFC magnetization at low temperatures. The isothermal remnant magnetization (IRM) was recorded using the following protocol: The sample was cooled in zero field to a specific temperature T_h below the maximum in the ZFC magnetization. At this temperature, the sample was kept for a specific wait time, t_w. After the wait time the magnetic field was applied (1 mT) and kept constant during the hold time, t_h. The magnetic field was cut to

Journal of Physics: Conference Series **521** (2014) 012012 doi:10.1088/1742-6596/521/1/012012

zero and the sample was left with an IRM. The sample was then, immediately after the field had reached zero, cooled at the maximum cooling rate of the MPMS (10 K/min) to a lowest temperature (10 K). When reaching this temperature, the $IRM(T)$ was recorded on heating the sample at a constant reheating rate (2 K/min) to a temperature above the glass transition temperature.

Figure 1 a) and b) shows the results of measurements using this method on an Ising spin glass and a dilute magnetic alloy $Cu(Mn)$ representing a system with Heisenberg spins. There is a striking difference in between the behavior of the two model systems on approaching T_h; the $Cu(Mn)$ $IRM(T)$ curve shows a pronounced maximum just before the expected rapid decrease to zero just above T_h, such an anomalous maximum does not appear in the $IRM(T)$ curve for the Ising spin glass. (The temperature scale in figure 1 a) and b) has been normalized to the magnetization in the FC at T_h).

The maghemite particles of this study were synthesized by thermal decomposition [7]. The particles are single domain and have a narrow size distribution with a mean diameter of 8 nm. The sample used in this study was prepared by pressing the (γ-Fe_2O_3) nanoparticles into a compact disc. The filling factor of the sample is about 67% which is close to the value corresponding to random close packing. The spin glass temperature T_g for the sample was determined to 140 K [5, 8]. A dilute superparamagnetic reference sample prepared from the same batch of particles had a blocking temperature of 36 K, indicating that the strongly enhanced random dipolar interaction of the dense sample governs the dynamics at higher temperatures [5, 7].

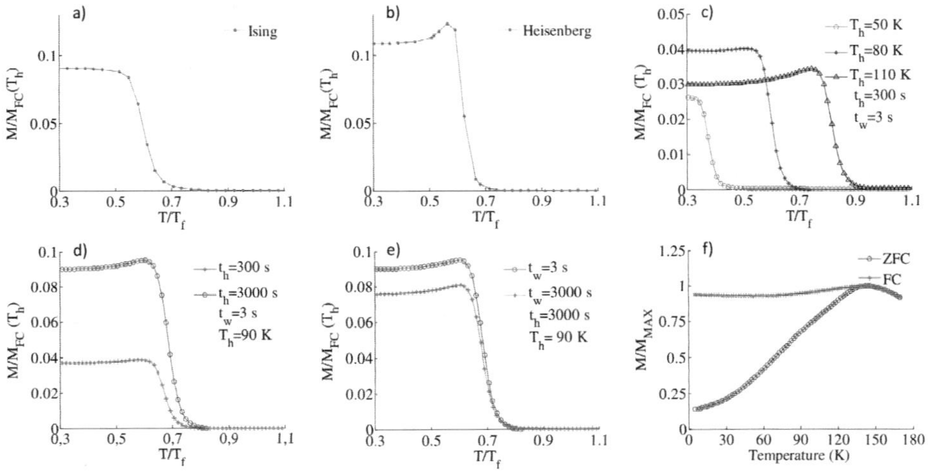

Figure 1. a) and b) show Typical $IRM(T)$ curves for an Ising ($Fe_{0.5}Mn_{0.5}TiO_3$) and a Heisenberg ($Cu(Mn)$) spin glass, adapted from Fig. 3 in ref. [6].) c) $IRM(T)$ for three different halting temperatures, T_h=50, 80 and 110 K. d) $IRM(T)$ at T_h=90 K with different t_h= 300 and 3000 s. e) $IRM(T)$ at T_h=90 K with different t_w= 3 s and 3000 s. f) ZFC and FC magnetization which have been normalized to the FC magnetization at 145 K. In a)-e), each $IRM(T)$ curve is normalized to the FC magnetization value at T_h. The applied field is in all experiments 1 mT.

3. Results and Discussion

Figure 1 shows the temperature dependence of the magnetization of the sample using the different protocols described above and applied fields of 1 mT. Figure 1 f) shows the ZFC

and FC magnetization curves as reference for the magnitude and temperature dependence of the magnetization. Figures 1 c-e) show different IRM-measurements where the magnetization is normalized to the value of the FC magnetization curve at T_h and the temperature is normalized to T_f. In the three experiments using different temperatures T_h, a pronounced peak is observed for the curve corresponding to the highest T_h, a weaker maximum for the intermediate T_h and for the lowest temperature no peak can be seen. Comparing these observations to those of the Ising and Heisenberg spin glasses of figure 1 a) and b), one is tempted to interpret the behavior as a crossover from essentially Ising to Heisenberg like super spins with increasing temperature. It can also be seen that the magnitude of the IRM for the curve corresponding to the middle halting temperature is larger than for both the lowest and the highest T_h. In figure 1 d), a plot at T_h/T_g=0.6 with two different values of t_h; 300 s and 3000 s and in figure 1 e) a plot using the same T_h and the same t_h=3000 s but different t_w = 3 s and 3000 s before the field was applied are shown. Both these figures confirm the expected behavior that the magnitude of IRM increases with t_h and that ageing slows down the relaxation, yielding a lower magnitude of the IRM for larger wait times.

Our experimental results are interpreted to indicate a crossover from Ising to Heisenberg character of the super spins of the particles on increasing temperature. This conclusion require certain caution: In atomic spin glasses, the spins are classified as being of Ising or Heisenberg character. The super spins of an super spin glass are on the other hand classical magnetic moments, i.e. the quantum mechanical concepts Ising and Heisenberg character are not applicable. However, at low temperature the magnetic moment of a particle is essentially directed along the magnetically easy axis in-between flips and may be considered as an Ising like super spin. At higher temperatures, the thermal fluctuations of the magnetic moment away from the easy direction become excessive in-between flips and the dimensional character of the super spin effectively crosses over to attain Heisenberg like character.

Acknowledgments

Financial support from the Swedish Research Council (VR) and the Göran Gustafsson foundation is acknowledged.

References

[1] Jönsson P E 2004 *Advances in Chemical Physics* ed Rice S A (John Wiley & Sons, Inc., Hoboken) pp 191–248
[2] Bedanta S and Kleemann W 2009 *Journal of Physics D: Applied Physics* **42** 013001
[3] Hansen M F, Jönsson P E, Nordblad P and Svedlindh P 2002 *Journal of Physics: Condensed Matter* **14** 4901
[4] Hiroi K, Komatsu K and Sato T 2011 *Phys. Rev. B* **83** 224423
[5] De Toro J A, Lee S S, Salazar D, Cheong J L, Normile P S, Muniz P, Riveiro J M, Hillenkamp M, Tournus F, Tamion A and Nordblad P 2013 *Applied Physics Letters* **102** 183104
[6] Mathieu R, Hudl M, Nordblad P, Tokunaga Y, Kaneko Y, Tokura Y, Katori H A and Ito A 2010 *Philosophical Magazine Letters* **90** 723–729
[7] De Toro J A *et al.* 2013 *The Journal of Physical Chemistry C* **117** 10213–10219
[8] Mathieu R, De Toro J A, Salazar D, Lee S S, Cheong J L and Nordblad P 2013 *Europhysics Letters* **102** 67002

Giant magnetoresistance in cluster-assembled nanostructures: a quantitative approach

Simón Oyarzún[1], Artur Domingues Tavares de Sá[2], Véronique Dupuis[1] and Matthias Hillenkamp[1]

1 Institut Lumière Matière, UMR5306 Université Lyon 1-CNRS, Université de Lyon, 69622 Villeurbanne cedex, France

2 Instituto de Física Gleb Wataghin, UNICAMP, CP 6165, Campinas, SP 13083-970, Brazil

E-mail: matthias.hillenkamp@univ-lyon1.fr

Abstract. In a recent publication [1] we have discussed qualitatively the applicability of commonly used assumptions and models in the description of giant magnetoresistance in granular media by comparing to well-defined cluster-assembled nanostructures of cobalt clusters embedded in copper thin films. In this article we present a quantitative analysis and discuss the relevant parameters appearing in the model.

1. Introduction

Spin-dependent transport in granular media is a field of contemporary research with high potential as well as many remaining open questions. We have, in a recent publication [1], described the fabrication and study of benchmark cluster-assembled nanostructures consisting of well-defined cobalt nanoparticles around 2.3 nm in diameter embedded in copper matrices. They were prepared from preformed gas-phase clusters following the low-energy cluster beam deposition technique [1, 2]. The thorough magnetic characterization based on the "triple fit" [3] permitted us to unambiguosly determine the median magnetic size, its dispersion as well as the effective anisotropy constant of the nanoparticles. Low temperature (2 K) remanence measurements furthermore allowed us to verify that inter-particle interactions are negligible at very low concentrations (0.5 at.% in this case), a crucial prerequisite for the commonly accepted description of giant magnetoresistance (GMR) in granular systems [4, 5].

Besides testing for the first time the model to describe GMR introduced by Zhang and Levy [6] with a truly superparamagnetic sample, we addressed the influence of spin-dependent scattering at the interface and within the volume of the nanoparticle. The literature is controversial [4, 7] and we could, using a parametrized formula, deduce that the stated discrimination between the two influences is highly questionable because of inter-particle interactions.

In order to better describe our data and continue testing the model, we decided to use the whole formula derived by Zhang and Levy and fit our data quantitatively. While this quantitative description is expected to give more insight in the underlying physics, it also introduces a considerable number of free parameters. We have seen before that the magnetic information contained in the room temperature magnetization cycle is highly ambiguous [3, 8], so we have to beware of the same ambiguity also in the magnetoresistance curves. We show in this article a first quantitative fit to our data and a limitation to the applicability of the model.

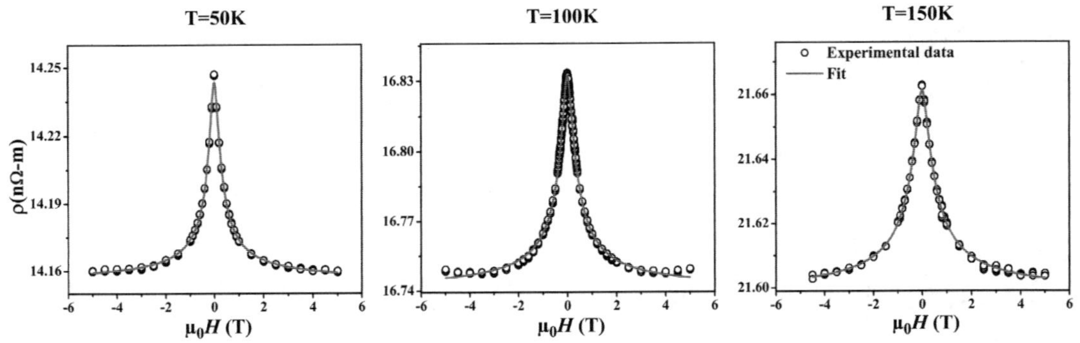

Figure 1. Experimental data (points) for the Co:Cu sample with 0.5 at.% taken at different temperatures and fits according to the model described in the text. The data can equally well be reproduced with only interface ($p_b = 0$) or only volume ($p_s = 0$) scattering. The curve for $T = 200$ K can be found in [1].

2. Discussion

The commonly used model to describe GMR in granular media was introduced by Zhang and Levy [6] and leads to the following expression for the field-dependent variation of the resistivity:

$$\frac{\Delta\rho}{\rho} = -\frac{\xi_1^2}{\xi_0^2} \tag{1}$$

where

$$\xi_0 = \frac{1-c}{\lambda_{nm}} + \frac{c}{\lambda_m}(1+p_b^2) + \frac{a_0(36\pi)^{1/3}c(1+p_s^2)}{\lambda_s}\frac{\int V_\alpha^{2/3}f(V_\alpha)dV_\alpha}{\int V_\alpha f(V_\alpha)dV_\alpha} \tag{2}$$

$$\xi_1 = \frac{2cp_b}{\lambda_m}\frac{\int V_\alpha f(V_\alpha)m_\alpha(V_\alpha)dV_\alpha}{\int V_\alpha f(V_\alpha)dV_\alpha} + \frac{a_0 2(36\pi)^{1/3}cp_s}{\lambda_s}\frac{\int V_\alpha^{2/3}f(V_\alpha)m_\alpha(V_\alpha)dV_\alpha}{\int V_\alpha f(V_\alpha)dV_\alpha}. \tag{3}$$

The parameters used are the mean free paths $\lambda_{nm,m,s}$ in the non-magnetic matrix, inside the magnetic grains and at the surface of the grain, respectively; p_s and p_b represent the ratios of the spin dependent to spin independent potentials for the interface and the volume of the nanoparticles. The nanoparticle concentration (molar fraction) is denoted by c, a_0 is the lattice constant of the nanoparticle material, $f(V_\alpha)$ the size distribution function and m_α the corresponding magnetic moment. Field and temperature mainly enter the formula through the Langevin description of the magnetization, the other parameters except λ_{nm} show only very weak variations with H and T in the superparamagnetic state (as discussed below).

Additional information can be obtained from the comparison with a pure Cu film of comparable thickness and prepared under identical conditions. The $\rho(T)$ curve nicely overlaps with the one measured for the sample with 0.5 at.% Co concentration. This directly shows that λ_{nm} is dominated by scattering events within the thin film, e.g. at grain boundaries, dislocations or the surfaces [9, 10].

Fig.1 shows fits according to the model at different temperatures. λ_{nm} at the respective temperature is in agreement with the pure Cu film within the error bars and the Co bulk value was taken for the lattice constant a_0. This leaves as free parameters the median size and dispersion of the magnetic size distribution, the mean free paths $\lambda_{m,s}$ and the ratios of the spin dependent to spin independent potentials p_s and p_b. At all temperatures shown here

	200 K		150 K		100 K		50 K	
	Interface	Volume	Interf.	Vol.	Interf.	Vol.	Interf.	Vol.
p_s	0.7±0.3	0	0.7	0	0.7	0	0.9	0
p_b	0	0.7±0.3	0	0.7	0	0.7	0	0.7
λ_s (nm)	1.9±0.9	3.8	1.4	3.0	1.5	6.3	3.2	3.4
λ_m (nm)	25±10	3.6±1.6	25	3.7	25	3.1	5.3	5.2
λ_{nm} (nm)	26±2		31		42		53	
D^m_{GMR} (nm)	2.4±0.3	2.4	2.4	2.4	2.3	2.3	2.0	2.0
w^m_{GMR} (nm)	0.2±0.1	0.2	0.2	0.2	0.2	0.2	0.2	0.2

Table 1. Parameters obtained from fits to GMR data at different temperatures. Typical errors are displayed for 200 K. The median diameter D^m_{GMR} and the diameter dispersion w^m_{GMR} are to be compared to $D^m_{TEM} = 2.3$ nm [1].

the experimental data are very nicely reproduced by the quantitative model. As for 200 K [1] we obtain identical fits when restricting the scattering to either the surface ($p_b = 0$) or the cluster volume ($p_s = 0$), no definite conclusion about the different contributions can be taken at the moment. It should be noted that even if p_s or p_b are set to zero, a finite term for the corresponding mean free path has to be assumed since both appear in the spin-independent scattering term ξ_0.

Looking at the derived numbers shows that above 100 K, when the whole cluster ensemble is in the superparamagnetic state, all the remaining free parameters have reasonable values, no obviously unphysical values are obtained. At 50 K, however, the parameters start to change, and as the temperature gets closer to $T_{max} = 13$ K [1] the median determined size starts to deviate from the correct value of 2.3 nm. The superparamagnetic assumption is no longer valid since a fraction of the ensemble is already blocked.

In conclusion we have shown that it is possible to obtain a quantitative description of the GMR curves at high temperatures using the Zhang and Levy model. In particular, for $T > 5T_{max}$, we obtain the same magnetic size distribution as in the magnetic characterization. Nevertheless, due to the high number of free parameters in the model, it is impossible for now to discriminate between the postulated contributions to spin-dependent scattering from the nanoparticle volume and surface. Further experiments with varying nanomagnet diameters are in progress to illustrate finite size effects in spintronics.

References

[1] Oyarzún S, Domingues Tavares de Sa A, Tuaillon-Combes J, Tamion A, Hillion A, Boisron O, Mosset A, Pellarin M, Dupuis V and Hillenkamp M 2013 *J. Nanopart. Res.* **15** 1968
[2] Perez A, Dupuis V, Tuaillon-Combes J, Bardotti L, Prevel B, Bernstein E, Mélinon P, Favre L, Hannour A and Jamet M 2005 *Adv. Eng. Mater.* **7** 475–485
[3] Tamion A, Hillenkamp M, Tournus F, Bonet E and Dupuis V 2009 *Appl. Phys. Lett.* **95** 062503
[4] Allia P, Knobel M, Tiberto P and Vinai F 1995 *Phys. Rev. B* **52** 15398–15411
[5] Batlle X and Labarta A 2002 *J. Phys. D: Appl. Phys.* **35** R15–R42
[6] Zhang S and Levy P M 1993 *J. Appl. Phys.* **73** 5315
[7] Rubin S, Holdenried M and Micklitz H 1998 *Eur. Phys. J. B* **5** 23–28
[8] Tamion A, Hillenkamp M, Tournus F, Bonet E and Dupuis V 2012 *Appl. Phys. Lett.* **100** 136102
[9] Henriquez R, Cancino S, Espinosa A, Flores M, Hoffmann T, Kremer G, Lisoni J G, Moraga L, Morales R, Oyarzun S, Suarez M A, Zúñiga A and Munoz R C 2010 *Phys. Rev. B* **82**(11) 113409
[10] Henriquez R, Moraga L, Kremer G, Flores M, Espinosa A and Munoz R C 2013 *Appl. Phys. Lett.* **102** 051608

Study of the heating characteristics and mechanisms of magnetic nanoparticles over a wide range of frequencies and amplitudes of an alternating magnetic field

A. Seki[1], E. Kita[1], D. Isaka[1], Y. Kikuchi[1], K.Z. Suzuki[1], A. Horiuchi[1], M. Kishimoto[1], H. Yanagihara[1], T. Oda[2], N. Ohkohchi[2], H. Ikehata[3] and I. Nagano[3]

[1] Institute of Applied Physics, University of Tsukuba, Tsukuba, Ibaraki 305-8573, Japan
[2] Department of Surgery, Advanced Biomedical Applications, Graduate School of Comprehensive Human Science, University of Tsukuba, Tsukuba 305-8575, Japan
[3] Department of Natural Sci. and Tech., Kanazawa University, Kanazawa 920-1192, Japan

E-mail: kita@bk.tsukuba.ac.jp

Abstract. An effective approach to gaining an understanding of the mechanism of heat generation for magnetic hyperthermia from nanomagnet suspensions is to perform heating tests over a wide frequency range. We constructed a heating test apparatus by using ferrite field cores with air gaps for low frequencies and solenoids for high frequencies. Magnetic field amplitudes up to 600 Oe (400 Oe) can be generated for frequencies lower than 500 kHz (800 kHz). Heat generation tests were performed for ferromagnetic nano-platelets and Co-doped Fe_3O_4 nanoparticles over a wide range of frequencies.

1. Introduction

In the biomedical field, application of nanomagnets has attracted attention especially for cancer therapy. For use in magnetic hyperthermia cancer therapy, magnetic nanoparticles are required to have strong heating power. In general, employing an alternating magnetic field (AMF) with high frequencies and amplitudes results in high heat generation; however, there is a limitation on the AMF, as it has a considerable effect on living matter due to eddy currents.[1] Therefore, effective use of the AMF is necessary and we have demonstrated the advantage of using ferromagnetic nanoparticles with cubic symmetry.[2] The major mechanisms of heat generation by magnetic nanoparticles, which include ferromagnetic hysteresis, superparamagnetic (SPM) relaxation (Néel relaxation), and Brownian relaxation, should be analyzed and combined to schieve effective use of AMF.

The heating characteristics of a magnetic nanoparticle suspension over a wide range of AFM frequencies and amplitudes are worth investigating in order to understand and differentiate between these heating mechanisms because the responses of the nanoparticles depend on each mechanism. In this paper, a high-performance magnetic field generator (up to 400–600 Oe) with a wide frequency range (50–800 kHz) was built, and the mechanism of heat generation in several different types of water-based nanoparticle suspensions was examined experimentally.

2. Magnetic field generator

An AMF was generated using series resonant LC circuits. A magnetic field, $H(t) = H_0 \sin(2\pi f_0 t)$, with an amplitude of H_0 was produced at a resonance frequency of f_0. Ferrite cores[3] and solenoids[4] were used as field generators. Commercial ferrite cores (TDK, PC-95) with a 40-mm inner diameter, a 60-nm outer diameter, and a 10-mm air gap were used for frequencies under 200 kHz, and Litz-wire windings with 10–20 turns were formed. For frequencies higher than 200 kHz, solenoid coils with inner diameters of 16 mm were used, and Litz wires (660 wires, each 0.04 mm in diameter) were used to make windings with 100 turns.

The circuit diagram is shown in Fig. 1. The sinusoidal wave from a signal generator was amplified by a 200 W power amplifier (Thamway, T145-5314A). To achieve better impedance matching, a transformer with a 12-turn primary coil and 1-turn secondary coil was used between the power supply and the field generator. The current flowing in the coil was monitored by a Hall-probe current sensor (U-RD, HSC-20-SC-A2.5) with a wide current range.

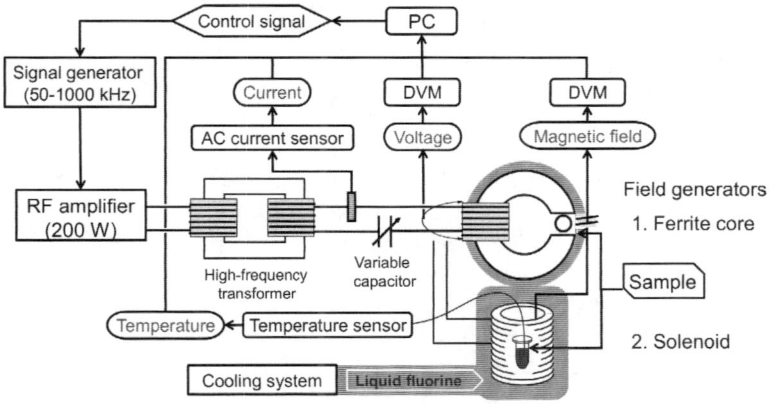

Figure 1. Block diagram of the AMF generator system.

The performance of the field generator system is shown in Fig. 2. The maximum amplitude of the AMF, H_0, decreased almost linearly with increasing operating frequency and was about 600 Oe for frequencies lower than 400 kHz and was 400 Oe for a frequency of 800 kHz. The maximum amplitude is limited by the Joule heating at the coil even though we used Litz wires. The solid line in the figure corresponds to a constant AMF frequencyamplitude product of $f_0 \cdot H_0 = 6.5 \times 10^7$ Hz·Oe; this value represents a recommended safety guideline.[1]

The heating characteristics were measured by setting water suspension samples and a fluorescent thermometer in the space where the AMF existed, and the specific loss power (SLP) was deduced from the initial change in the time variation of the temperature.

3. Heating characteristics of nanoparticles

Two kinds of samples were examined for the heat generation test: dimple-contained nanoparticles (DINP)[5] and Co-doped Fe_3O_4 nanoparticles(SCF)[2]. The DINP have an oval shape with approximate dimensions of 80 x 40 x 10 nm for the long and short axes and the thickness, respectively. The coercive force of a dried sample is around 150 Oe, mostly exhibiting bulk ferromagnetic properties. The SCF particles have a particle size of 20 nm and a coercive force of 145 Oe in a dry state. Water–based suspensions of these particles were made with a particle concentration of 0.9 and 3.7 wt% for DINP and SCF, respectively. These two particles have almost the same coercive force; however, the heating characteristics are different (see Fig. 3).

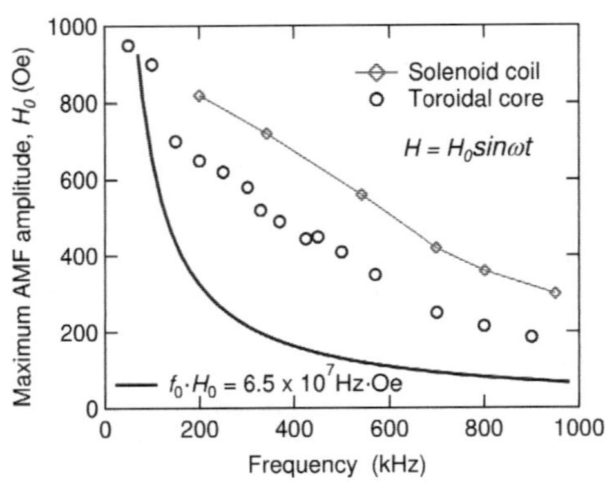

Figure 2. Performance of the AMF generator with a wide frequency range. The solid line shows the safety limit corresponding to a constant value of $f_0 \cdot H_0$.

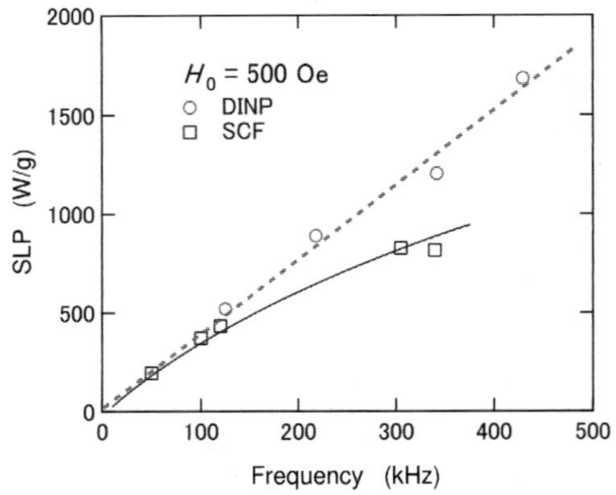

Figure 3. Frequency dependence of the specific loss power for ferromagnetic nano-platelets (DINP)[5] and Co-doped Fe_3O_4 nanoparticles (SCF).

For typical ferromagnetic particles with sufficient coercive force, the major heating mechanisms are hysteresis loss and/or Brownian relaxation. For the case of the ferromagnetic nano-platelets (DINP), because the particle size and the hydrodynamic radius V_H are large, the characteristic time for Brownian relaxation, $\tau_B = 3\eta V_H/k_B T$ and $f_B = 1/\tau_B$ are estimated to be 2×10^{-4} s and 5 kHz, respectively, which is much lower than the test frequencies. For the present case, the viscosity of the suspension, η, is 1 mPa·s. Thus, it was concluded that the heating mechanism was governed by the magnetic hysteresis loss. The power loss Q is simply proportional to the hysteresis area S and the frequency, f_0, such that $Q = f_0 S$. The experimental data[5] clearly shows the linear dependence on the AFM frequency and supports the idea that the ferromagnetic hysteretic mechanism is responsible for the heat generation. This mechanism might be stabilized as a result of the dipolar interaction between ferromagnetic particles even though the concentration is 0.9 wt%

On the other hand, because the SCF sample has a small diameter and a spherical shape, the hydrodynamic radius V_H is small, and the characteristic time for Brownian relaxation, τ_B, is estimated to be 5×10^{-6}s suggesting that the deviation of the SLP curve from a linear relationship with f_0 may take place at around 200 kHz. This tendency can be seen in Fig. 3 and Co-doped Fe_3O_4 nanoparticles (SCF) can be explained by the combination of magnetic hysteresis and Brownian relaxation, respectively.

Acknowledgements
This work was supported by Grants-in-Aid for Scientific Research (Grant No. 23300185, and 23300362) from the Ministry of Education, Culture, Sports, Science and Technology of Japan.

References

[1] Hergt R and Andrä W 2007 *Magnetism in Medicine* Hoboken, NJ, Wiley, p. 550
[2] Kita E, Oda T, Kayano K, Sato T, Minagawa M, Yanagihara H, Kishimoto M, Mitsumata C, Hashimoto S, Yamada K and Ohkohchi N 2010 *J. Phys. D: Appl. Phys.* **43** 474011
[3] Lacroix L-M, Carrey J and Respaud M B 2008 *Rev. Sci. Instrum.* **79** 093909
[4] Fortin J-P, Wilhelm C, Servais J, Menager C, Bacri J-C and Gazeau F 2007 *J. Am. Chem. Soc.* **129** 2628
[5] Kishimoto M, Yanagihara H and Kita E 2013 *IEEE Trans. Magn.* **49** 4756

8th International Conference on Fine Particle Magnetism (ICFPM2013) IOP Publishing
Journal of Physics: Conference Series **521** (2014) 012015 doi:10.1088/1742-6596/521/1/012015

Computational study on superparamagnetic hyperthermia with biocompatible SPIONs to destroy the cancer cells

C. Caizer

Department of Physics, West University of Timisoara, Timisoara, Romania.

e-mail: ccaizer@physics.uvt.ro

Abstract. Superparamagnetic hyperthermia (SPMHT) appears nowadays as the most promising method of the future, non-invasive and with low toxicity, for destroys the cancer cells through the magnetic relaxation in superparamagnetic nanoparticles. In our research we focused on finding the *optimal* conditions using a 3D computational study to obtain a *maximum* specific absorption rate (SAR) by the magnetic relaxation in Fe_3O_4 and γ-Fe_2O_3 superparamagnetic iron oxide nanoparticles (SPIONs), which give the most pronounced SAR and with low toxicity on cells. The effect of the diameter of the nanoparticles, frequency and amplitude of external alternating magnetic field and the thickness of biological coating of nanoparticles in the case of their encapsulation in biocompatible membranes, like liposomes (Ls) and cyclodextrins (CDs), on Néel-Brown magnetic relaxation and maximum SAR, are presented and discussed in this paper, within the biological admitted limit.

1. Introduction

Superparamagnetic hyperthermia (SPMHT) [1], obtained as a result of magnetic relaxation in superparamagnetic iron oxide nanoparticles (SPIONs), is a very suitable method for destroying cancer cells [2], by increasing the temperature to *42-43 °C* in the targeted tissues. This method is non-invasive and apparently without toxicity when nanoparticles are encapsulated in biocompatible membranes, such as liposomes (Ls) or cyclodextrins (CDs), which are used today as a possible nanocarriers in drug delivery [3,4]. One of the most important issues in the use of SPMHT is finding the *optimum* conditions for obtaining a *maximum* specific absorption rate (SAR). Several research studies have been conducted on the effect of the nanoparticles' diameter upon the magnetic hyperthermia [5,6]. However, finding the *optimal physical* conditions for obtaining a *maximum* SAR, within the accepted biological limit [7] is still an open issue, which must be clarified before moving on to the next stage, of *in vitro* or *in vivo* studies and, finally, to the clinical studies. The success of the *in vitro/in vivo* magnetic hyperthermia highly depends on the results from the first stage. In this regard, we conducted a 3D computational study of SAR, considering simultaneously the diameter of the nanoparticles (D), the frequency (f) and amplitude of the magnetic field (H) in the case of SPIONs of Fe_3O_4 and γ-Fe_2O_3, when they are encapsulated in Ls or bioconjugated with CDs (using CDs in SPMHT is a new strategy of ours). Our 3D study allows us to *rapidly* find the *optimum* values of the *D, f, H* parameters, the viscosity of the dispersion environment of the bionanocapsules, etc., by simply modifying their values in the calculation program, thus optimizing the superparamagnetic hyperthermia, by simulating real conditions. The study focused on finding the optimum diameter of the nanoparticles, which is a critical parameter, and on the way in which the thickness of the biological membrane of the Ls and of the layer determined by the CDs at the surface of the biocompatible nanoparticles, when dispersed in water, influence the contribution of the Neel-Brown relaxation processes in obtaining the maximum SAR. Furthermore, for this study, we took into consideration a very small volume fraction of the nanoparticles, of only *1.7 vol%*, which allows the maintenance of the superparamagnetic behavior of nanoparticles [8], without the appearance of the hysteresis loop caused by the interactions between nanoparticles [9], and the application of the Neel-Brown relaxation theory; in static conditions (where the measuring time is *~100 s* [10]), in the lack of the interactions between nanoparticles, and as a result of the reduced magnetic anisotropy of the Fe_3O_4 and γ-Fe_2O_3 nanoparticles, the superparamagnetic (SPM) behavior is maintained for rather large diameters (*<20 nm*), because the Neel-Brown relaxation times (in the order of nano - microseconds) are much lower than the measuring time (tens – hundreds of seconds). Consequently, considering the Neel-Brown magnetic relaxation, we aimed to find the *optimum* conditions in which the *highest maximum* SAR is *obtained* for the nanoparticles of Fe_3O_4 and γ-Fe_2O_3, when they are encapsulated in Ls or bioconjugated with CDs, *without exceeding the biological limit* [7].

2. Theory

Volumetric power dissipated in superparamagnetic nanoparticles in alternating harmonic magnetic field, with frequency f and amplitude H [1], can be written based on the static magnetic susceptibility χ_0 and magnetic relaxation time τ [11]: $P = \mu_0 \pi f \dfrac{2\pi f \tau}{1+(2\pi f \tau)^2} \chi_0 H^2$. In low magnetic fields, χ_0 can be approximated by the initial magnetic susceptibility $\chi_i = \varepsilon \mu_0 \pi M_s^2 D^3 / 18 k_B T$, deduced from the Langevin function [12], for the approximation of the spherical nanoparticles, and the relaxation time is $\tau = \tau_N \tau_B / (\tau_N + \tau_B)$, with the following components: $\tau_N = \tau_0 \exp\left(\pi K D^3 / 6 k_B T\right)$, for the Néel relaxation time, and $\tau_B = 3\pi \eta D_h^3 / 6 k_B T$, for the Brown relaxation time. In the above mentioned formulas, M_s is the spontaneous magnetization, D is the magnetic (mean) diameter, ε is the (volume) magnetic packing fraction of nanoparticles, μ_0 is the permeability of vacuum, k_B is the Boltzmann constant, T is the room temperature, K is the magnetic anisotropy constant, τ_0 is the time constant ($\sim 10^{-9}$ *s*), η is the viscosity coefficient and D_h is the hydrodynamic diameter of nanoparticles: $D_h = D + 2d$, d being the biological coating thickness. Specific absorption rate (SAR), under adiabatic conditions, can be expressed in terms of power dissipation, expressed in *W/g*, by the formula: $SAR = P / \rho$, where ρ is the material density.

3. Results and discussions

SAR was studied in 3D for the monodispersed Fe_3O_4 and γ-Fe_2O_3 nanoparticles, which are suitable for magnetic hyperthermia. SAR was also studied when the nanoparticles are encapsulated in Ls or bioconjugated with CDs, while considering the hydrodynamic diameter of the nanobiocapsules (D_h). In these conditions, we aimed to find the most suitable nanoparticles, regarding the diameter (optimal diameter in the range of *1- 30 nm*), depending on the frequency and amplitude of the external magnetic field, which are leading to the *highest maximum* SAR, without exceeding the biological limit. Data used in calculations, according to the formulas in Section 2, are given in Table 1.

Tabel 1: *Characteristic observables of nanoparticles and alternating magnetic field parameters*

Samples	M_s *(kA/m)*	K *(kJ/m³)*	ρ *(g/cm³)*	ε	d_{Ls} [3] *(nm)*	d_{CDs} [14] *(nm)*	η *(kgm⁻¹s⁻¹)*	f *(kHz)*	H *(kA/m)*
Fe_3O_4 [13]	477	11	5.24	0.017	35	0.8	7×10^{-4}	100-1000	10-20
γ-Fe_2O_3 [1]	416	4.6	5.20	0.017	-	-	-	100-1000	10

Due to the very low *(0.017)* volume magnetic packing fraction of nanoparticles (ε) and to the thickness of the biological coating of the nanoparticle, the nanoparticles are well isolated from one another and, therefore, the interactions between them (Van der Waals and dipole-dipole) may be neglected [8,15].

In the case of Fe_3O_4 nanoparticles (figure 1), for the *10 kH/m* amplitude of the magnetic field, it has been found that the maximum SAR can be obtained if the magnetic diameter of the nanoparticles is *~15-17 nm*, depending on the frequency *(100 - 1000 kHz)*; at lower frequencies the diameter has a higher value. Highest maximum for SAR is *~27 W/g* and it can be obtained at the diameter of *~15 nm* and frequency of *1 MHz*.

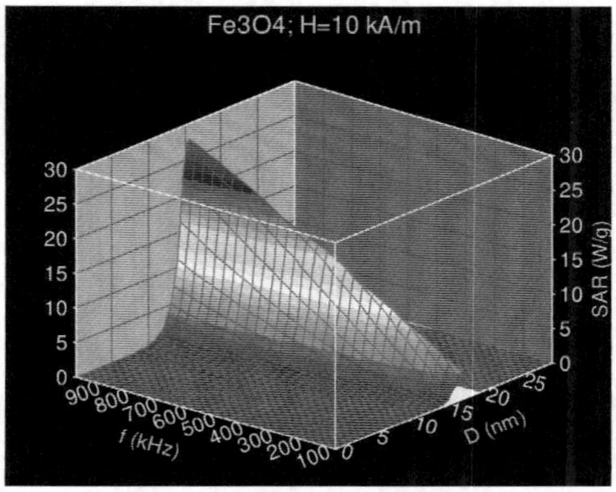

Figure 1. (colour online) 3D diagram of SAR in the case of Fe_3O_4 nanoparticles.

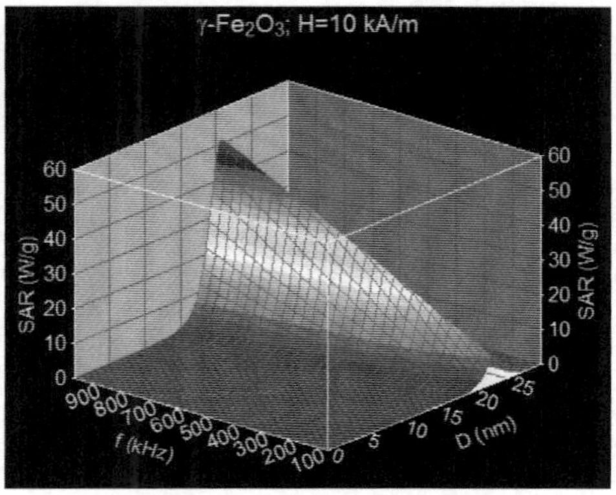

Figure 2. (colour online) 3D diagram of SAR in the case of γ-Fe_2O_3 nanoparticles.

In the case of γ-Fe_2O_3 nanoparticles (figure 2), the SAR increase, reaching the highest value of *50 W/g*, under the same field conditions and frequency, but at considerably larger diameter of nanoparticles, i.e. *~20.5 nm*. However, although a higher maximum SAR it is obtain for γ-Fe_2O_3 nanoparticles, the high value of the nanoparticles' diameter determines their behaviour in external magnetic field to not be superparamagnetic (Néel magnetic relaxation time increases rapidly with diameter between *19-25 nm*), and thus appear a small hysteresis loop. Consequently, the formula for SAR cannot be used in this case. For this reason, Fe_3O_4 nanoparticles are more suited than those of γ-Fe_2O_3 in *superparamagnetic* hyperthermia. Other authors have shown that, in the case of γ-Fe_2O_3 nanoparticles, the maximum specific loss power (SLP) is obtained for smaller diameters, namely *14 nm* [5]. This is probably due to the value of the magnetic anisotropy constant higher than the known one [1,16]; a higher anisotropy constant leads to a maximum of the SLP at a smaller diameter.

Having in view this result, we studied the SAR for Fe_3O_4 nanoparticles when they are made biocompatible in order to be used in practice. Further on, we consider two cases: (i) Fe_3O_4 encapsulated in Ls and (ii) Fe_3O_4 bioconjugated with CDs, strategy currently used in research on targeted therapy. When the nanoparticles are encapsulated in Ls (i), Brown relaxation time becomes much higher than the Néel relaxation time (τ_B $(167\times10^3\ ns)$ >> τ_N $(122\ ns)$), and the relaxation time will be: $\tau = \tau_N/(1+\tau_N/\tau_B) \cong \tau_N$ (see chapter 2). In this case the SAR diagram obtained is similar to that in figure 1, these being determined only by the Néel dissipation processes.

The results for Fe_3O_4-CDs are shown in figure 3 for two values of the magnetic field, considering biological accessibility limit, $Hf \leq 5\times10^9\ Am^{-1}Hz$ [7]. In this case, two magnetic relaxation processes contribute to the SAR, both Néel and Brown, for magnetic diameters greater than *15 nm*. Néel relaxation processes become dominant with the increasing frequency to the upper limit at the *~15.5-17 nm*. At lower frequencies, such as *150 kHz*, contributions to the SAR of the Néel and Brown relaxation processes become comparable and at the frequency of *100 kHz*, Brown relaxation prevail, especially in

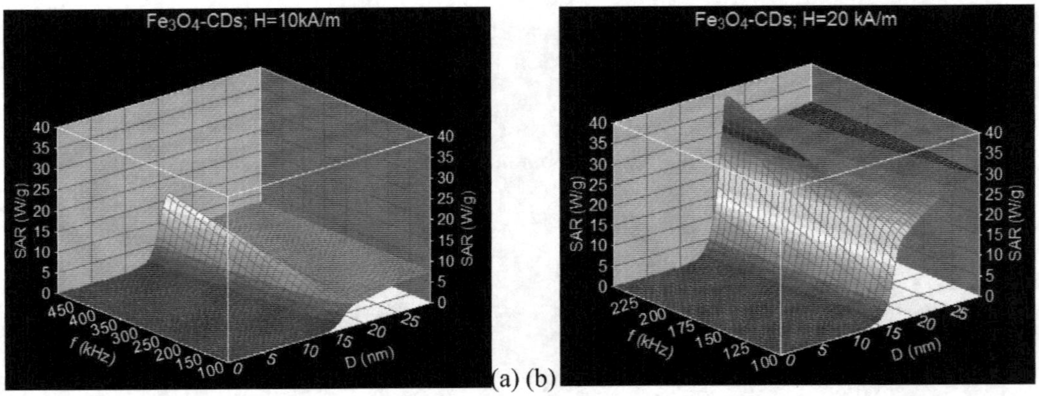

Figure 3. (colour online) 3D diagram of SAR until the biological limit in the case of Fe_3O_4 nanoparticles bioconjugated with CDs, for magnetic field amplitude of (a) 10 kA/m and (b) 20 kA/m.

large diameters. When the magnetic field reaches to *20 kA/m* (figure 3(b)), the highest maximum of SAR increases significantly from *~15 W/g* to about *~38 W/g*, at the biological limit, for the *17 nm* diameter. In this case, the SAR still remains high (*~30 W/g*) for *D > 22 nm*, due to Brown relaxation.

4. Conclusions

The 3D study allowed us to optimize superparamagnetic hyperthermia, in order to apply it in practice: by finding the most suitable bionanoparticles, as material and size, and also as magnetic field parameters (*f* and *H*), in order to obtain the highest SAR within the biological accepted limit.

References

[1] Rosensweig R E 2002 *J. Magn. Magn. Mater.* 252 370
[2] Tanaka K, Ito A, Kobayashi T, Kawamura T, Shimada S, Matsumoto K, Saida T and Honda H 2005 *Inter. J. Cancer* 116 624
[3] Caizer C, Hadaruga N, Hadaruga D, Tanasie G and Vlazan P *7th Int. Conf. on Inorganic Materials* 12 – 14 September 2010, Biarritz, France
[4] Huang S H 2008 *Adv. Drug Deliv. Rev.* 60 1167
[5] Gazeau F, Lévy M and Wilhelm C 2008 *Nanomedicine* 3 831
[6] Habib A H, Ondek C L, Chaudhary P, Bockstaller M R and McHenry M E 2008 *J. Appl. Phys.* 103 07A307
[7] Hergt R and Dutz S 2007 *J. Magn. Magn. Mater.* 311 187
[8] Caizer C 2003 *J. Phys.: Condens. Matter* 15 765
[9] Bakoglidis K D, Simeonidis K, Sakellari D, Stefanou G and Angelakeris M 2012 *IEEE Trans. Magn.* 48 1320
[10] Néel L 1955 *Adv. Phys.* 4 191
[11] Shliomis M 1974 *Sov. Phys. Uspekhi* 17 153
[12] Jacobs I S and Bean C P 1963 *Magnetism* vol III, ed G T Rado and H Suhl (New York: Academic Press)
[13] Smit J and Wijin H P J 1961 *Les Ferrites* (Paris: Bibl. Tech. Philips)
[14] Li J and Loh X J 2008 *Adv. Drug Deliv. Rev.* 60 1000
[15] Chantrell R W, Walmsley N S, Gore J and Maylin M 1999 *J. Magn. Magn. Mater.* 118 169
[16] Mørup S 1983 *J. Magn. Magn. Mater.* 37 39

AUTHOR INDEX

Institute of Physics
Dirac House, Temple Back
Bristol BS1 6BE UK

CURRAN ASSOCIATES INC
proceedings.com